OXFORD SERIES ON ADVANCED MANUFACTURING

SERIES EDITORS

J. R. CROOKALL
MILTON C. SHAW

OXFORD SERIES ON ADVANCED MANUFACTURING

1. William T. Harris *Chemical milling: the technology of cutting materials by etching* (1976)
2. Bernard Crossland *Explosive welding of metals and its applications* (1982)
3. Milton C. Shaw *Metal cutting principles* (1984)
4. Shiro Kobayashi *Metal forming and the finite element method*
5. Norio Taniguchi, Masayuki Ikeda, Iwao Miyamoto, and Toshiyuki Miyazaki *Energy-beam processing of materials: advanced manufacturing using various energy sources* (1989)
6. Nam P. Suh *Axiomatic approach to manufacturing process and product optimization*
7. N. Logothetis and H. P. Wynn *Quality by design: experimental design, off-line quality control, and Taguchi's contributions*
8. John L. Burbidge *Production flow analysis for planning group technology*

ENERGY-BEAM PROCESSING OF MATERIALS

Advanced Manufacturing Using Various Energy Sources

Norio Taniguchi

Professor at Tokyo Science University

WITH CONTRIBUTIONS BY
Masayuki Ikeda
Iwao Miyamoto
Toshiyuki Miyazaki

CLARENDON PRESS · OXFORD
1989

Oxford University Press, Walton Street, Oxford OX2 6DP
Oxford New York Toronto
Delhi Bombay Calcutta Madras Karachi
Petaling Jaya Singapore Hong Kong Tokyo
Nairobi Dar es Salaam Cape Town
Melbourne Auckland
and associated companies in
Berlin Ibadan

Oxford is a trade mark of Oxford University Press

Published in the United States
by Oxford University Press, New York

© Norio Taniguchi, 1989

All rights reserved. No part of this publication may be reproduced,
stored in a retrieval system, or transmitted, in any form or by any means,
electronic, mechanical, photocopying, recording, or otherwise, without
the prior permission of Oxford University Press

British Library Cataloguing in Publication Data
Taniguchi, Norio
Energy-beam processing of materials.
1. Materials. Processing. Applications of electron beams &
lasers. Electrons. Beams. Applications in processing of materials
I. Title
6201'12
ISBN 0-19-859005-9

Library of Congress Cataloging in Publication Data
Taniguchi, Norio. 1912–
Energy-beam processing of materials: advanced manufacturing using
various energy sources / Norio Taniguchi.
p. cm.—(Oxford series on advanced manufacturing; 5)
Includes bibliographies and index.
1. Manufacturing processes. 2. Electron beams—Industrial
applications. 3. Ion bombardment—Industrial applications.
4. Photon beams—Industrial applications. I. Title. II. Series.
TS183.T36 1989   670.42—dc 19   88-38656
ISBN 0-19-859005-9

Typeset by Macmillan India Ltd, Bangalore-25

Printed in Great Britain by
Biddles Ltd,
Guildford & King's Lynn

PREFACE

This book is intended to provide an introduction to advanced manufacturing processes using new energy sources, special emphasis being placed on the energy-beam processing of materials.

The principal author, Norio Taniguchi, Professor at Tokyo Science University, started his research on the mechanisms of machining processes of hard and brittle materials (such as jewels, precious stones, optical glasses, quartz crystals, silicon crystals, and ceramics) in 1940 just after the onset of the Second World War. Later he directed his research to the non-conventional processing of hard and brittle materials.

His first area of interest was ultrasonic machining; later he turned to the various kinds of energy-beam processes such as electric discharge machining and microwave, electron-beam, and laser-beam processing.

His present research is on ultra-precision processing using ion-beam energy; ion-beam processing may be used successfully for ultra-precision work and the making of fine parts. The practical manufacture of components of microelectronic devices and electro-optical devices requires processing resolution of the order of 0.01 μm (10 nm) and accuracy of less than 1 nm (defined by least machining dimension) is required. However, nowadays, micro-mechanical components require the fineness of a few μm.

The co-author of Chapter 2 (Photon-beam processing) is Dr Masayuki Ikeda, Principal Scientist of the National Electric Technical Laboratory in Japan. His doctoral thesis was concerned with the lapping of non-metallic crystals, but he later changed his research area to the advanced technology of laser-beam processing. He is now in charge of the development of laser-beam processing in the flexible manufacturing system complex, sponsored by MITI (Ministry of International Trade and Industry).

The co-author of Chapter 3 (Electron-beam processing) and Appendix A is Dr Toshiyuki Miyazaki, Professor at the Japan Governmental Institute of Vocational Training. His doctoral thesis was concerned with laser-beam processing, but he is now engaged mainly in research into the mechanisms of electron-beam processing.

The co-author of Chapter 4 (Ion-beam processing) and Appendix B is Dr Iwao Miyamoto, Lecturer at Tokyo Science University. His doctoral thesis was concerned with ion-beam sputter machining for ultra-precision mechanical components.

The co-authors are amongst those engaged in the most promising research on energy-beam processing in Japan.

The author would like to express his sincere thanks to Professor J. R. Crookall for his careful work in checking and correcting the drafts, especially the English. Furthermore he would like to add his thanks to Professor M. C. Shaw for his kind advice, to Professor P. A. McKeown for his helpful remarks on the drafts, and also to Professor A. Kobayashi for his continuing help with the research. Finally, he would like to express his thanks to the staff of Oxford University Press for help in publishing this book.

Tokyo Science University   N. T.

CONTENTS

List of Contributors ix

1. **Introduction** (Norio Taniguchi) 1

   Basic concept of energy-beam processing   Concept of specific processing energy of materials   Estimating the threshold specific processing energy of materials   Power density in energy-beam processing   Photon and related energy-beam processing   Electron and related energy-beam processing   Ion and related energy-beam processing   Atom or molecule energy-beam processing   Plasma energy-beam processing   Chemical and electrochemical reactant energy-beam processing   Micro-fabrication techniques using energy-beam processing of materials   Summary   References

2. **Photon-beam processing** (Masayuki Ikeda) 62

   Introduction   Fundamentals of laser-beam processing   Equipment for laser-beam processing   Laser-beam machining   Welding and soldering   Surface modification   New applications   Conclusions   References

3. **Electron-beam processing** (Toshiyuki Miyazaki) 138

   Introduction   Electron-beam machining   Electron-beam welding   Electron-beam lithography   Other kinds of electron-beam process   References

4. **Ion-beam processing** (Iwao Miyamoto) 200

   Introduction   Ion-beam removal   Ion-beam deposition   Ion-beam surface treatment   Other applications of ion-beam processing   References

Appendix A. Temperature analysis of thermal energy processing by electron beam and laser beam (Toshiyuki Miyazaki)   276

Basic equation of thermal energy processing based on heat conduction theory   Temperature rise in a semi-infinite solid due to a surface heat source   Temperature rise in a plate due to a surface heat source   Consideration of the process based on electron penetration   Effects of pulsed beam on the heated zone   Effect of variation of thermal properties with temperature   Latent heat of fusion and/or vaporization   Computer simulation of thermal energy-beam drilling   Temperature analysis around the melted zone in welding   Temperature analysis in the melted zone in welding   References

Appendix B. Monte Carlo computer simulation of ion-beam processing (Iwao Miyamoto)   294

Introduction   Monte Carlo simulation of scattering ions within target materials   Modifications to the simulation model for collision of ions of lower energy   Adaptation of the Kinchin–Pease model   Results of Monte Carlo simulation   Conclusions   References

Index   310

CONTRIBUTORS

Masayuki Ikeda, Principal Scientist at the National Electric Technical Laboratory, Japan

Iwao Miyamoto, Lecturer at Tokyo Science University

Toshiyuki Miyazaki, Professor at the Japan Governmental Institute of Vocational Training

Norio Taniguchi, Professor at Tokyo Science University

1
INTRODUCTION

1.1 Basic concept of energy-beam processing

1.1.1 Needs for energy-beam tools

To date, machining processes using solid cutting tools are the principal finishing processes for materials. Hence they are referred to as conventional manufacturing processes. At the time of World War II, materials which are difficult to cut, such as hard and tough special steels and alloys, hard and brittle new engineering ceramics, and soft but tenacious new engineering plastics, were becoming more widely used as machined components in industrial products. Consequently, technical problems of how to cut or machine these materials economically and accurately have arisen.

To solve these problems, many excellent new cutting tools, abrasives, and grinding wheels, as well as more effective machining processes and machine tools, have been developed. However, despite these improvements, the most serious problems of tool wear in machining process have not yet been solved.

Lately, new machining processes have been developed which use solid machining tools, aided by electrical, chemical, and electrochemical processes, as shown in Table 1.1. These processes are very useful in solving tool wear problems and they herald the introduction of energy-beam processing using new energy sources, without any kind of solid machining tool.

1.1.2 Types of energy-beam processing

In energy-beam processing, all solid machining tools, such as cutting tools, milling cutters, abrasives, and grinding wheels, with their accompanying problems of wear, are dispensed with; only energy beams are used. The processing energy, in various forms, is supplied directly to the working point in the form of a beam. The energy beam is defined as a directional flux of extremely small energetic particles, such as photons, electrons, ions, plasma, or chemically and electrochemically reactive atoms.

The energy-beam processing techniques which have been developed may be classified into six categories, as follows:

(i) photon energy-beam processing
(ii) electron energy-beam processing

TABLE 1.1
Mechanical processing of materials aided by a wide variety of energy sources

1. Cutting and related processes
   high-temperature cutting (electric, flame, or plasma), low-temperature cutting (liquid CO_2), etc.
2. Grinding and related processes
   ultrasonic honing and grinding, electrolytic grinding, electric-discharge grinding, chemical grinding, etc.
3. Abrasive working and related processes
   electromagnetic or electrostatic abrasive lapping and polishing, chemical abrasive lapping and polishing, ultrasonic abrasive machining, etc.
4. Plastic flow forming
   hot pressing and forming, dieless hot drawing and forming, electromagnetic forming, explosive forming, centrifugal free forming, forming by induction and dielectric heating, drawing aided by ultrasonic vibration, etc.
5. Joining, fixing, and related processes
   thermal diffusion press-fixing, chemical reaction press-fixing, press-fixing aided by high-frequency electric heating, soldering aided by ultrasonic vibration, percussion welding aided by direct electric discharge, etc.
6. Powder forming
   ultrafine powder sinter-forming, and hot isostatic press forming, aided by electric heating, pressing and packing aided by ultrasonic vibration, etc.
7. Surface treatment
   peening aided by electric heating and vibrations

(iii) ion energy-beam processing
(iv) atom or molecule energy-beam processing
(v) plasma energy-beam processing
(vi) chemical and electrochemical reactant energy-beam processing.

The processing systems outlined above are shown schematically in Fig. 1.1, and types of energy-beam processing are listed in Table 1.2. A summary of recent research and development on these processing techniques is given in a paper by Taniguchi [1].

This book provides detailed explanations of photon-beam, electron-beam, and ion-beam processing, but plasma-beam, chemical and electrochemical, and other beam techniques are not treated to any great extent, because these are dealt with in detail in other specialist books.

Photon (laser) beam
(focused or broad beam)

Electron beam
(focused or broad beam)

Electric discharge
current beam
(tool-guided beam)

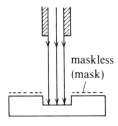

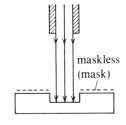

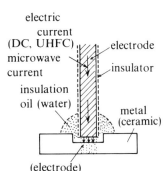

(air)
thermal processing
(photon reactive)
(removal, consolidaion)

(vacuum)
thermal processing
(electron reactive)
(removal, consolidation)

(insulation oil)
thermal processing
microwave current for
dielectrics (removal)

Ion beam
(focused or
broad beam)

Reactive ion beam
(cold plasma beam)
(broad beam)

Atomic/molecular
beam (cold molecular,
and hot molecular)
(broad beam)

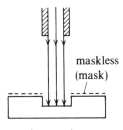

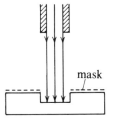

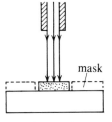

(vacuum)
dynamic processing
(ion reactive)
(removal, deposition)

(vacuum)
chemical and
sputter etching
(dynamic)
(removal, deposition)

(vacuum)
thermal and dynamic
processing
(deposition, diffusion)

FIG. 1.1

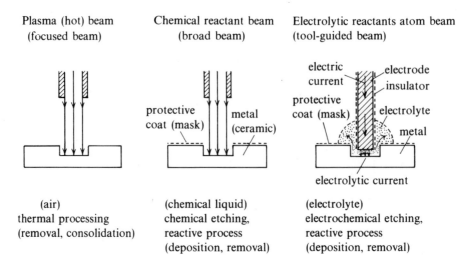

Fig. 1.1. Models showing energy-beam processing (removal and deposition)

1.2 Concept of specific processing energy of materials

1.2.1 Mechanical processing and energy-beam processing

In conventional machining processes, a mechanical breaking stress, generated by the cutting edge of a solid machining tool, plays an important role in removing material from the workpiece. In this case, a chip of considerable size is removed mechanically, as a result of fracture or failure originating from defects in the material, such as cracks, voids, and grain boundaries. Hence, this kind of removal process is called mechanical processing.

On the other hand, energy-beam processing is performed by removing material by increasing the internal energy of atoms in the workpiece; consequently, chip removal is effected atom by atom, or atom cluster by atom cluster. Therefore, energy-beam processing must be treated as a problem involving the processing energy. In practice, the technique is performed as a thermal, reactive, or dynamic process causing removal, joining, or slip at the atomic scale. Hence it is called 'atomic-scale processing' [2–4].

1.2.2 Internal energy of atoms, lattice bonding energy, and surface barrier energy

According to the theory of the thermodynamics of solids, the internal free energy U of an atom of a solid substance consists of the Gibbs free energy G (J/atom), thermal energy H (J/atom), internal elastic strain energy E_e (J/atom),

TABLE 1.2
Types of energy-beam processing of materials

1. Photon-beam and related energy-beam processing

Processing mechanism	Removal/separation	Joining/consolidation	Forming	Surface treatment
Light beam (ultra violet to infrared)	*Thermal processes:* Laser (YAG, CO_2, etc.)/heat-ray machining	Laser/heat-ray welding	Laser/heat-ray surface polishing	Laser/heat-ray surface hardening
	Reactive processes: Photoetching, exposing (aided by chemical etching)			Polymerization/depolymerization
Radiation (γ-ray, X-ray, and SOR)	*Reactive processes:* Radiation exposure/etching			Polymerization/depolymerization

2. Electron-beam and related energy-beam processing

Electron beam	*Thermal processes (bombardment):* Electron-beam machining	Electron-beam welding and evaporated deposition	Electron-beam surface polishing	Electron-beam surface hardening, annealing

TABLE 1.2
Types of energy-beam processing of materials

2. Electron-beam and related energy-beam processing

Processing mechanism	Removal/separation	Joining/consolidation	Forming	Surface treatment
	Reactive processes (excitation):			*Reactive processes (excitation):*
	Electron-beam exposure/etching			Electron-beam polymerization/depolymerization
	Thermal processes (bombardment, and resistive, inductive, and dielectric heating):			
Electric-discharge beam (direct current for metal/ultra-high frequency current for dielectrics)	Electric-discharge machining (die-sinking, piercing, wire-cutting)	Electric-discharge percussion welding	Electric-discharge surface polishing	Electric-discharge surface hardening

3. Ion-beam and related energy-beam processing

Ion beam	*Dynamic processes:*			
	Ion-sputter etching, ion-assisted chemical etching	Ion implantation, ion plating, ion deposition, ion-sputter deposition	Ion rubbing	Ion nitration/ion mixing
	Reactive processes:			
	Reactive-ion etching, cold-plasma etching	Reactive-ion plating		Reactive-ion surface finishing

4. Atom or molecule energy-beam processing

	Dynamic and thermal processes:			
Cold and hot atom/molecule beam	Atom/molecule diffused etching	Atom/molecule deposition, spray crystal epitaxial growth	Atom/molecule dynamic finishing	Atom/molecule thermal diffusion

5. Plasma energy-beam processing

	Thermal processes:			
Hot plasma beam	Plasma jet beam machining	Plasma arc beam welding (TIG, MIG), plasma jet beam fusion, coating, joining	Plasma arc beam flow polishing	Plasma jet/arc surface hardening

6. Chemical and electrochemical reactant energy-beam processing

	Chemical processes:			
Chemical beam	Chemical machining/etching, chemico-mechanical etching, photo-etching	Chemical plating/casting, chemical reactive/vapour deposition, chemical cementing, synthetic crystal growth	Chemico-mechanical flow polishing	Chemical reactive surface finishing
	Electrochemical processes:			
Electrochemical beam	Electrochemical machining	Electrochemical plating/casting		Electrochemical reactive finishing

and linear kinetic energy E_k (J/atom):

$$U = G + H + E_e + E_k \tag{1.1}$$

where: G is the potential energy for chemical or electrochemical decomposition or activation; H is the thermal energy due to atomic vibration around the lattice site; E_e is expressed by $\int_0^x f\,dx$, where f is a potential force and x is the displacement of the atom from its lattice site; and E_k is the linear kinetic energy.

The Helmholtz free energy, F (J/atom), is defined as

$$F = G + E_e \tag{1.2}$$

corresponding to the internal free energy U minus the thermal energy, i.e. U when $H = 0$ (0 K) and $E_k = 0$ (steady state). The Helmholtz free energy F varies with E_e, i.e. the atomic displacement x from the lattice site, as shown in Fig. 1.2, and becomes a minimum at the lattice site ($x = 0$). In other words, the Helmholtz free energy of an atom at its lattice site corresponds to the internal free energy which holds the atom at that site. Consequently, the Gibbs free

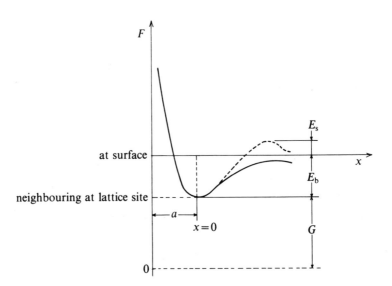

FIG. 1.2. Helmholtz free energy vs. displacement of atom. $F =$ Helmholtz free energy of atom $= G + E_e$ (internal energy U, minus H and E_k); $x =$ displacement of atom from atomic lattice site; $a =$ atomic lattice length; $E_b =$ lattice bonding energy; $E_s =$ surface barrier potential energy; $U =$ internal free energy of atom, $U = G + H + E_e + E_k$; $G =$ Gibbs free energy of atom; $H =$ thermal energy of atom; $E_e =$ elastic strain energy of atom; $E_k =$ linear kinetic energy of atom

energy corresponds to the minimum Helmholtz free energy of the atom at its lattice site.

To displace the atom from its lattice site and remove it from the surface of the workpiece, the internal free energy of the atom must be increased beyond the level of Helmholtz free energy at the surface of the workpiece. This is larger than the minimum Helmholtz free energy by the lattice bonding energy, E_b (J/atom), which corresponds to $[E_e]_\infty = \int_0^\infty f\,dx$, as shown in Fig. 1.2.

The free energy necessary to remove the atom from the surface of the workpiece is obtained by increasing the amplitude of thermal vibration of the atom with a projected photon or electron, or by supplying kinetic energy E_k by ion bombardment, or by similar means.

In practice, the increase in internal energy or processing energy required to remove an atom from the surface of the workpiece must be greater than the lattice bonding energy E_b, and also there must be a surplus of surface barrier energy E_s, (J/atom), which corresponds to the surface barrier potential, as shown in Fig. 1.2.

Therefore, the threshold specific processing energy δ (J/m³) is defined as the minimum additional internal energy required per unit volume of material to be removed. That is,

$$\delta = (E_b + E_s) N_A / v \quad (\text{J/m}^3) \tag{1.3}$$

where N_A is Avogadro's number, 6.022×10^{23} atom/mol, and v is the molar volume of the work material, m³/mol. The value of δ is very important in energy-beam processing.

1.2.3 Processing unit size and threshold specific processing energy

Machining processes which use solid tools are performed mainly by mechanical energy or increased internal elastic strain energy, transferred through the cutting tool edge or abrasive grain tip. However, in energy-beam processes, the processing energy is supplied directly to the workpiece, by means of the energy beam. Accordingly, in removal or separation processes, the threshold specific processing energy δ (J/m³) for removal should first be considered. However, this threshold energy varies considerably with the processing unit size ε (m), where ε corresponds approximately to the chip size in cutting or grinding. In energy-beam processing, ε corresponds macroscopically to the working size of the area over which processing energy is supplied, but microscopically to the chip size, that is, the atom or atomic cluster in thermal, reactive, and ionic bombardment processes.

The reason that the threshold specific processing energy δ varies so greatly is that the mechanisms of fracture and failure in processes such as removal, joining, and deformation depend on the kinds of active defects existing in the region of the processing unit size, as shown in Fig. 1.3.

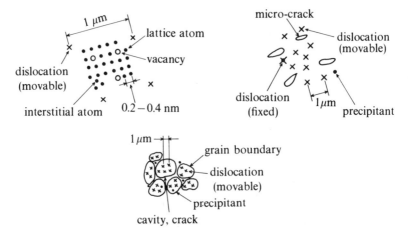

Fig. 1.3. Models of defect distribution in materials

1.2.4 Variation of fracture or failure mechanism with processing unit size [5]

(1) Fracture and failure on the atomic lattice scale

The smallest processing unit size may be considered to be an atomic lattice distance of 0.3 to 0.4 nm, because atomic structure itself may act as a defect causing a fracture or failure in this domain. In other words, in the processing unit size range $\varepsilon = 10^{-5}$ to 10^{-3} μm, removal, joining, and deformation at the atomic scale are based on direct separation, consolidation, and slip between atoms, as a result of their increased internal energy.

In practice, the threshold specific processing energy δ for the atomic lattice range is in the region of 10^5 to 10^4 MJ/m^3 for chemical and electrochemical processing, thermal melting, and evaporative processing, and 10^6 to 10^5 MJ/m^3 for ion sputter processing, as shown in Table 1.3.

(2) Fracture and failure on the point-defect scale

The next-largest processing unit size may be considered to be about $\varepsilon = 10^{-3}$ to 10^{-1} μm, which corresponds to the distribution interval of point defects in crystals, such as vacancies and interstitial atoms, as shown in Fig. 1.3. In this point-defect range, the threshold specific processing energy δ becomes 10^4 to 10^3 MJ/m^3 for atomic-cluster energy processing, as shown in Table 1.3.

Furthermore, fracture and failure occur as a result of atomic shear slip caused by elastic yield shear, and tensile fracture caused by cleavage between atomic clusters. Mirror surface finishing and carbon fibre cutting are examples of this kind of atomic-cluster processing.

(3) Fracture and failure on the dislocation–microcrack scale

The next-largest processing unit size, $\varepsilon = 10^{-5}$ to 10 μm, corresponds to the distribution interval of movable dislocations, which is about 1 μm in a crystal grain of a polycrystalline metallic material. A fixed dislocation or precipitant exists more frequently in metallic materials, but these defects are assumed to act as fixed lattice atoms.

In metal manufacture, cutting and plastic working are based on the slip caused by movable dislocations, because this type of plastic deformation occurs more readily than atomic elastic shear slip or failure. That is to say, the Peierls force on the movable dislocation is quite small compared with ordinary elastic shear deformation.

In inorganic crystals such as single silicon crystals, there exist few dislocations, but many microcracks with the same distribution range. Hence processing on this scale is called dislocation–microcrack mechanical processing. Furthermore, amorphous ceramics, such as glass and fused silica (SiO_2), show both an irregularity called microphase, and microcracks which have nearly the same distribution range. Therefore, processing making use of microcracks is to be expected in ceramic crystals.

Finally, organic polymers have irregularities of several μm, and fracture or failure occurs at a very low working energy density, as a result of slip between layers of polymer molecules which are bonded by the very weak van der Waals forces. Therefore, this may be considered as a process on the dislocation– microcrack scale.

However, in the micro-cutting of polymers generating a chip of submicrometre size, it becomes necessary to supply considerable specific processing energy, of the order of 10^4 to 10^3 MJ/m^3, which corresponds to processing in the point-defect range. This is because, in order to break the fibres of polymers directly, it is necessary to initiate tensile fracture based on point defects.

(4) Fracture and failure on the grain–void scale

In the case of larger processing unit sizes, > 0.1 mm, which corresponds to the mean grain size of various materials, fracture or failure based on defects such as cracks, cavities, and grain boundaries occurs. Therefore, the necessary threshold specific processing energy δ becomes fairly small, about 10 MJ/m^3.

1.3 Estimating the threshold specific processing energy

1.3.1 Threshold for mechanical processing

(1) Threshold specific processing energy for initiation of plastic slip failure

Plastic slip occurs as a result of dislocation defects in ductile materials, such as steel or copper. The cutting shear stresses, corresponding to the threshold specific slip energy, vary with the processing unit size, i.e. depth of cut or chip thickness, as shown in Fig. 1.4, and described by Beck et al. [6].

TABLE 1.3
Threshold specific processing energy δ (J/m^3) (mainly for removal)

Processing unit size ε (m)	Atomic lattice range 10^{-11}–10^{-9}	Point defect range 10^{-9}–10^{-7}	Dislocation–microcrack range 10^{-7}–10^{-5}	Grain–void range 10^{-5}–10^{-3}	Remarks (other processes)
Defects/heterogeneity	Atom/molecule	Point defects/ (vacancy/ interstitial atom)	Movable dislocation/ micro-crack	Crack/cavity/ grain boundary	
Processing mechanism					
Chemical/electro-chemical reactive decomposition	10^{11}–10^{10} (atomic)				Deposition (electro-chemical plating, reactive photoelec-tric)
Brittle/tensile cleavage/fracture		10^{10}–10^9 (atomic tensile breaking)	10^9–10^8 (micro-cracking)	10^8–10^7 (brittle cracking)	Removal (glass, ceramics etc.)
Shear failure slip/ dislocation slip		10^{10}–10^9 (atomic shear slipping)	10^9–10^8 (disloc-ation slip)	10^8–10^7 (plastic deformation)	Slipping (ion rubbing)/deform (ductile breakdown)
		10^{10}–10^9 (atomic cluster)	($\omega = 10^6$–10^1) (ω = specific volumetric stock removal energy, J/m^3)		

Melting off/diffusion (thermal)	10^{11}–10^{10} (atomic)	10^{10}–10^{9} (atomic cluster)	Deposition (ion nitrating/mixing)
Evaporation (thermal)	10^{11}–10^{10} (atomic)	10^{10}–10^{9} (atomic cluster)	Deposition (vacuum evaporation, ion sputter deposition, etc.)
Ion sputtering off/atomic dynamic separation	10^{12}–10^{11} (atomic)	10^{11}–10^{10} (atomic cluster)	Consolidation (ion implantation ion deposition)
	Atomic-scale processing	Atomic cluster processing	Dislocation–micro-crack mechanical processes	Grain size mechanical processing

For ductile materials

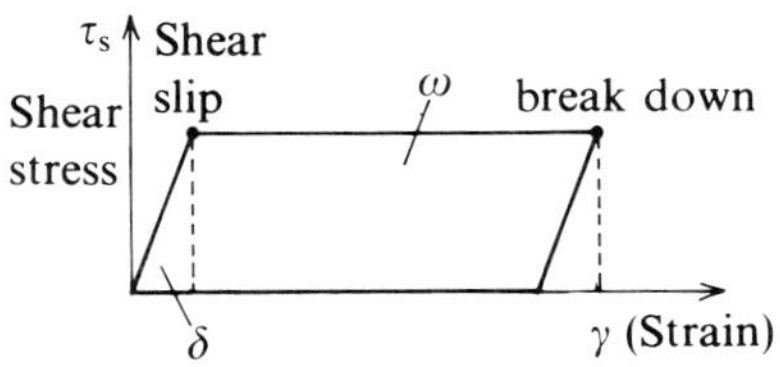

τ_s Shear stress — Shear slip — ω break down — δ — γ (Strain)

For brittle materials

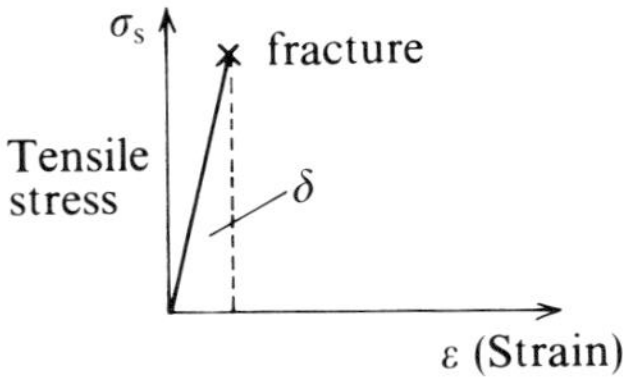

σ_s Tensile stress — × fracture — δ — ε (Strain)

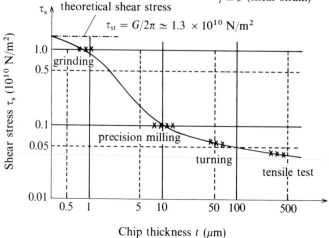

FIG. 1.4. Micro-cutting and increase of shear stress (after Beck et al., Trans. ASME **74** (1952), 1)

The cutting shear stress τ_s (N/m²), acting on the cutting edge of the abrasive grain in the fine grinding of steel, is about 1.0×10^4 MN/m², which corresponds to the energy for a chip 1 μm in thickness and is nearly the theoretical shear strength of steel τ_{st}, obtained from the following equation:

$$\tau_{st} = G/2\pi = 8.16 \times 10^{10}/2\pi = 1.3 \times 10^4 \text{ MN/m}^2 \qquad (1.4)$$

where G is the rigidity modulus, 8.16×10^4 MN/m² for steel. Accordingly, the threshold specific shearing energy, corresponding to a processing unit size in the smaller dislocation range ($\varepsilon = 1$ μm), can be estimated as follows:

$$\delta_{st} = \tau_{st}^2/2G = 1.03 \times 10^3 \text{ MJ/m}^3 \qquad (1.5)$$

The cutting shear stress for the larger chip thickness of 10 μm in precision milling, corresponding to a processing unit size in the larger dislocation range ($\varepsilon = 10$ μm), is obtained from Fig. 1.4 as follows:

$$\tau_s = 0.1 \times 10^4 \text{ MN/m}^2$$

and the corresponding threshold specific processing energy is

$$\delta_s = 6.0 \text{ MJ/m}^3$$

For larger chip sizes of 50 to 100 μm, corresponding to a processing unit in the grain size range ($\varepsilon \geqslant 0.1$ mm),

$$\tau_s = 0.05 \times 10^4 \text{ MN/m}^2$$

and

$$\delta_s = 1.5 \text{ MJ/m}^3$$

However, in the case of stock removal processes using larger plastic slip or strain γ, the specific volumetric stock removal energy ω (J/m^3) should be considered, as shown at the foot of Table 1.3. The value can be calculated as the shearing energy, expressed by the following formula:

$$\omega = \tau_s \gamma.$$

For the cutting of steel the value is about 10^3 MJ/m^3, and for the polishing of glass, 10^6 MJ/m^3 for $\gamma \approx 3$.

When precision mechanical processing such as micro-cutting or mirror cutting of steel becomes necessary, the chip thickness must be in the submicrometre region and, as a result, wear of the tool edge becomes too great for high precision, and tool materials such as high-speed steel, tungsten carbide, and diamond cannot be used successfully. The reason that wear of the tool edge becomes so great is that mechanical adhesion, abrasion, and chemical erosion become extremely large at the very high temperatures and pressures encountered at such a small processing unit size, owing to the larger threshold specific processing energy.

(2) Threshold specific processing energy for initiation of tensile crack fracture

Tensile fracture occurs mainly as a result of microcracks in brittle materials such as glass, ruby, and ceramics. The threshold value is determined by the existing probability of microcracks, i.e. the microcrack range given in Table 1.3.

Unlike ductile materials, the specific volumetric stock removal energy ω of brittle materials is equal to the threshold specific processing energy δ, because a microcrack in brittle material can grow by itself without surplus energy and form the cleavage surface of the chip.

1.3.2 Threshold for energy-beam processing

(1) Threshold for the atomic lattice range

In the case of a processing unit size $\varepsilon \leqslant 10^{-2}$ to 1 nm, processing is performed at the atomic scale, i.e. atom by atom. That is, the processing energy corresponds to the energy required for each atom to move against the potential field around its lattice site. Therefore, the threshold specific processing energy on the atomic lattice scale must be greater than the sum of the specific lattice bonding energy U_b (J/m^3) and the specific surface barrier energy U_s (J/m³), which correspond to the atomic bonding energy E_b (J/atom) and the atomic surface barrier energy E_s (J/atom) respectively.

Values of specific bonding energy U_b are shown in Table 1.4, as given by Plendle [7]. For example, the value of U_b for steel is about 2.6×10^3 MJ/m³ and the corresponding value of E_b is about 1.6×10^{-20} J/atom. Values of U_b for tensile fracture of Al_2O_3, SiC, and diamond I are 6.2×10^5, 1.38×10^6, and 5.64×10^6 MJ/m³ respectively.

TABLE 1.4

Specific volumetric lattice bonding energy U_b and atomic bonding energy E_b

Material	U_b (J/m³)	E_b (J/atom)	Remarks
Fe	2.6×10^9	1.6×10^{-20}	
	(1.03×10^9)	(8×10^{-21})	For shear slip
Al_2O_3	6.2×10^{11}	5.26×10^{-18}	
Si	7.5×10^{11}	1.59×10^{-17}	
SiC	1.38×10^{12}	1.1×10^{-17}	
CBN	2.09×10^{12}	1.7×10^{-17}	(Cubic boron nitride)
B_4C	2.26×10^{12}	1.8×10^{-17}	
Diamond I	5.64×10^{12}	4.5×10^{-17}	Abundant N_2 (natural)
Diamond II	1.02×10^{13}	8.2×10^{-17}	N_2-free

U_b: Plendle, J. N. *Phys. Rev.* **123** (1961), 4; **125** (1962), 3
E_b: the author's estimate
For reference, σ_s: surface tension to helium
        Fe 1.74 J/m²
        SiO_2 0.27 J/m²

The value of the atomic surface barrier energy E_s can be calculated from the surface energy density σ_s (J/m²) for each material. For example, the surface energy density of steel in an atmosphere of helium gas is about 1.72 J/m². Therefore, estimating from the mean atomic cross-sectional area, 4×10^{-20} m², the necessary atomic surface barrier energy E_s is about 6.88×10^{-20} J/atom, which is nearly equal to the lattice bonding energy, E_b, of steel.

Moreover, the atom which is being removed has to be supplied with surplus energy to escape from the surface of the workpiece. This energy may be assumed to be of the order of 10 eV (1.6×10^{-19} J) at atmospheric pressure and about 1 eV in a higher vacuum. As a result, the threshold specific processing energy δ (J/m³) must be considered to be several times the value of the specific lattice bonding energy U_b. In general, to obtain a practical removal rate, a sufficiently larger specific processing energy must be supplied, as described in a later section.

Of course, in chemical or electrochemical decomposition processes, i.e. chemical or electrochemical etching, the separation or removal energy of the atom is given directly by the difference between the Gibbs free energies of the reacting material and of the reaction products, augmented by the activation energy due to thermal vibration, which is dealt with in the section on chemical reactant energy-beam processing.

In thermal removal processes such as melting or evaporation, the separation energy of the atom is the thermal vibration energy itself, and in ion sputtering processes, the separation energy is supplied by the impact of the bombarding energetic ion.

(2) Threshold for the point-defect range

In the case of a processing unit $\varepsilon = 1$ to 10^2 nm, for atomic cluster energy processing, the energy necessary to remove unit volume of material, atom cluster by atom cluster, against the potential field around each atomic cluster must be considered, as shown in Table 1.3. The threshold specific processing energy for chemical and electrochemical processing, brittle fracture, shear failure, slip melting, evaporation, and sputtering, etc., is of the order of 10^4 to 10^3 MJ/m^3, and is smaller than for the atomic-lattice range.

(3) Threshold for other types of energy-beam processing

Consolidation and deformation on the atomic and atom-cluster scale have nearly the same processing energy as does separation, because joining and displacement processes are the inverse of removal processes.

1.4 Power density in energy-beam processing

1.4.1 Power density requirements

Taking processing on the atomic or the atomic-lattice scale as an example, the threshold specific processing energy δ must be at least the sum of the specific lattice bonding energy U_b and the specific surface barrier energy U_s, as mentioned in the previous section; the values are of the order of 10^6 to 10^4 MJ/m^3, as seen in Table 1.3.

For chemical and electrochemical decomposition, thermal melting, diffusion, and evaporation processing, the values of δ are of the order of 10^5 to 10^4 MJ/m^3, but for ion sputtering, δ is estimated to be of the order of 10^6 to 10^5 MJ/m^3, because a considerable excess of energy is necessary for the removed atom to escape through the surrounding medium.

Therefore, an important technical problem in energy-beam processing is that of how to concentrate the processing energy on the atoms at the working point and how to raise the internal energy of the atoms to the necessary higher

TABLE 1.5
Input power density and machining speed in energy-beam removal processes

Processing mechanism	Macroscopic power density (mean) P_m (W/m^2)	Microscopic power density (atomic scale) P_a (W/m^2)	Machining speed (macroscopic) v (m/s)	Remarks
Chemical and electrochemical decomposition	10^4–10^5	10^{11}–10^{12} (atomic)	10^{-7}–10^{-6}	
Brittle/tensile fracture	10^8–10^9	10^{11}–10^{12} (abrasive)	Widely variable	Abrasive blasting
Shear failure/dislocation slip	10^8–10^9 (grinding) 10^6–10^7 (mirror cutting)	10^{11}–10^{12} (abrasive) 10^9–10^{10} (chip)	Widely variable	Grinding, mirror cutting
Melting off/diffusion	10^9–10^{12} (continuous electron or laser beam) 10^8–10^9 (high temp. plasma)	10^{10}–10^{11} (pulsed electron, laser, or electric discharge beam)	10^{-1}–10^0 (for peak pulsed beam)	Electron, laser beam, elec. discharge, and plasma beam processing
Evaporation	10^{10}–10^{11} (continuous laser beam)	10^{11}–10^{12} (pulsed laser beam)	10^{-2}–10^{-1} (for peak pulsed beam)	Electron and laser beam processing
Atomic sputtering/atomic elastic emission	10^5–10^6	10^{12}–10^{13} (atomic/molecular)	10^{-8}–10^{-7}	Ion beam processing elastic emission processing
	For energy-beam projection area	For processing unit size or bit pulse	$v = P_m/\delta$ δ = threshold specific processing energy (J/m^3)	

level. Table 1.5 shows the practical macroscopic energy-beam power density and the microscopic power density.

1.4.2 Diffusion of internal energy in solid materials

As a result of the impact of a processing energy beam, the level of internal free energy of atoms in the workpiece increases, owing to their increased thermal energy and elastic strain energy, as shown in eqn (1.1). Of course, the Gibbs free energy itself is not changed in this process but becomes effective only for chemical and electrochemical processing, as will be described later.

The excess internal energy in the form of thermal vibration and elastic strain energy is transferred to neighbouring atoms through the interatomic action of atomic bonding potential energy, or elastic strain energy, and atomic thermal vibration energy. Accordingly, the excess internal energy above the Gibbs free energy is dissipated in accordance with the inherent transfer rate of the material. In other words, the rate of transfer of thermal energy is determined by the temperature, specific heat capacity, density, and thermal conductivity of the workpiece, and also by the area of the workpiece onto which the energy beam is projected. This will be dealt with later.

It is easily seen, therefore, that to maintain a greatly elevated level of internal energy of the atoms, the rate of supply of processing energy by the beam must equal the rate of dissipation of internal energy in the workpiece, which is determined by the inherent thermal constants of the material and also by the processing conditions. These phenomena should also be considered in the processes of consolidation and deformation.

In practice, even though the area over which the beam is projected is very small, the working temperature must be sufficiently high to cause melting or evaporation and, as a result, the rate of dissipation of energy from the working area becomes comparatively large. Consequently, the input power density of the energy beam must be considerable, as discussed in the next section.

1.4.3 Analysis of temperature in energy-beam processing

(1) Basic treatment of temperature analysis

The necessity for a high power density in the energy beam can be demonstrated by the heat conduction theory of Carslaw and Jaeger [8]. In laser- and electron-beam processing, which will be discussed later, thermal action seems to predominate. Therefore, it is important to calculate the temperature distribution and change in the workpiece.

Consider a thermal energy beam with an energy input rate, or power, Q (W), as shown in Fig. 1.5(a). The beam has a circular cross-section with uniform distribution of thermal flow. The calculation assumes that the energy

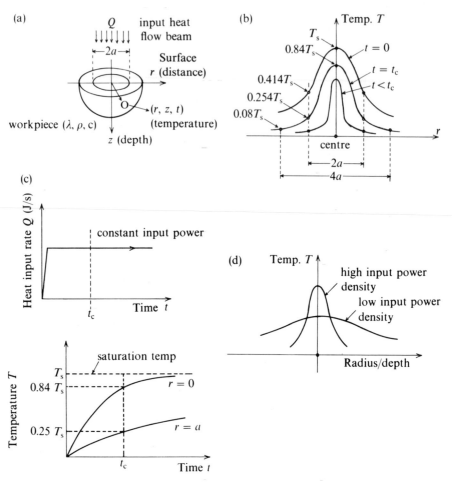

FIG. 1.5. Input of thermal energy, and temperature distribution and change. (a) Models of heating. $Q=$ energy input rate; $\lambda=$ conductivity; $\rho=$ density; $c=$ specific heat capacity; $t=$ time duration; $a=$ radius of projection area of beam. (b) Temperature distribution and gradient. $T_s=$ saturation temperature, $Q/(\pi\lambda a)=qa/\lambda$; $q=$ input power density, $Q/\pi a^2$; $\varphi_c=$ temp. gradient $=(0.84-0.254)T_s/a=0.6q/\lambda$. (c) Temperature change with constant input power. $t_c=$ characteristic response time, $\pi a^2/\kappa$; $\kappa=$ thermal diffusivity, $\lambda/\rho c$. (d) Variation of temperature distribution due to input power density

input rate is constant and that all energy supplied is converted to heat, or the thermal component of the internal free energy at the surface of the workpiece.

A detailed analysis will be given in Chapter 3 and Appendix A, but here the distribution and change of temperature in a semi-infinite workpiece are shown graphically. (Fig. 1.5(b), (c), and (d)) [9].

The most important factor in this temperature analysis is the saturation temperature T_s (K), due to a constant rate of input of thermal energy, Q (W), at the centre of the area of the workpiece surface onto which the thermal energy beam is projected. This is given by

$$T_s = Q/(\pi \lambda a) = qa/\lambda \tag{1.6}$$

where λ is the thermal conductivity of the workpiece, J/(m s K), a is the radius of the circular projection area, m, and q is the input thermal power density, W/m².

The other important factor is the characteristic response time t_c (s), as shown in Fig. 1.5(c), which is the time interval before the temperature at the centre of the projection area rises to $0.84 T_s$, and is determined as follows:

$$t_c = \pi a^2/\kappa \quad (t_c = a^2/\kappa \text{ in Appendix A}) \tag{1.7}$$

where κ is the thermal diffusivity $= \lambda/\rho c$, m²/s, ρ is the mass density, kg/m³, and c is the specific heat capacity, J/(kg K).

Another factor, the temperature gradient around the centre of the projection area at time t_c, is denoted φ_c (K/m) and is used as a reference for temperature distribution. It is taken as the ratio of the temperature difference between the centre and the periphery of the projection area at time t_c to the radius at the periphery:

$$\varphi_c = (0.84 - 0.254)\, T_s/a = 0.6 q/\lambda. \tag{1.8}$$

From this temperature gradient φ_c, the change in temperature distribution with input thermal power density q (W/m²) is shown conceptually in Fig. 1.5(d). It is readily seen that a beam of high input power density is necessary to obtain a steep temperature gradient, which is important for the piercing of small holes by direct energy-beam processing. On the other hand, to obtain the gentle temperature gradient required for surface treatment, etc., it is necessary to use a beam of lower input power density.

(2) Example of temperature analysis

As an illustration, take a carbon steel (0.1 per cent C) of thermal conductivity $\lambda = 0.458 \times 10^2$ W/(m K), mass density $\rho = 7.85 \times 10^3$ kg/m³, and specific heat capacity $c = 4.94 \times 10^2$ J/(kg K), with a projection area of radius $a = 0.1$ mm and input thermal power $Q = 50$ W. The input thermal power density q is thus 1.6×10^3 MW/m² and as a result, the saturation temperature $T_s = 3.48 \times 10^3$ K (above room temperature), and the characteristic response time $t_c = 2.66 \times 10^{-3}$ s, using eqns. (1.6) and (1.7) respectively.

Conversely, the constant input power density q necessary to elevate the temperature of the centre of the projection area to the melting point of steel, $T_m = 1800$ K, may be calculated using eqn. (1.6) and letting $T_s = T_m$. The result is $q = 8.24 \times 10^2$ MW/m².

In laser-beam or electron-beam processing using a constant pulsed step input, the input power density necessary for elevating the temperature of the periphery of the projection area to the melting point of steel, at the characteristic response time of steel under these conditions, $t_c = 2.66 \times 10^{-3}$ s, is about 3.25×10^3 MW/m^2. To reach the vaporization temperature, 3000 K, under the same conditions, the input power density required is 1.10×10^4 MW/m^2.

To estimate the heat-affected area in thermal energy-beam processing, it is necessary to calculate the temperature gradient around the centre of the projection area, φ_c, at the characteristic response time t_c under the above processing conditions. For an input power density sufficient for melting, i.e. $q = 3.25 \times 10^3$ MW/m^2 and $t_c = 2.6 \times 10^{-3}$ s,

$$\varphi_c = 4.25 \times 10^7 \text{ K/m}$$

TABLE 1.6
Temperature characteristics for thermal processing

	Diamond	Tungsten W	Beryllium oxide BeO	Copper Cu	Molybdenum Mo	Silicon Si
Thermal conductivity λ (J/(msK))	1.38–1.88 × 10²	1.69 × 10²	2.09 × 10²	3.89 × 10²	1.38 × 10²	1.13 × 10²
Specific heat c (J/(kgK))	5.0 × 10²	1.6 × 10²	10 × 10²	3.8 × 10²	3.1 × 10²	7.6 × 10²
Thermal diffusivity κ (m²/s)	9.5 × 10⁻⁵	5.30 × 10⁻⁵	7.3 × 10⁻⁵	1.14 × 10⁻⁴	4.91 × 10⁻⁵	6.33 × 10⁻⁵
Mass density ρ (kg/m³)	3.5 × 10³	19.3 × 10³	2.86 × 10³	8.94 × 10³	9.01 × 10³	2.33 × 10³
Characteristic response time t_c (s)	3.31 × 10⁻⁴	5.93 × 10⁻⁴	0.43 × 10⁻⁴	2.75 × 10⁻⁴	6.4 × 10⁻⁴	4.96 × 10⁻⁴
Saturation temperature T_s (K)	1.23 × 10³	1.21 × 10³	1.04 × 10³	6.81 × 10²	1.43 × 10³	1.69 × 10³
Melting point T_m (K)	(4000)	3650	2800	1360	2850	1690
Vaporization temperature T_v (K)	–	6200	–	3130	5073	–
Latent heat of melting h_m (J/m³)	–	3.56 × 10⁹	–	1.87 × 10⁹	2.64 × 10⁹	–
Latent heat of vaporization h_v (J/m³)	–	–	–	4.39 × 10¹⁰	–	–
Corrected melting temperature T_{mc} (K)	–	4761	–	1907	3790	–
Specific input thermal power of melting/unit radius, Q_{msp} W/m	8.12 × 10⁶	7.28 × 10⁶	6.99 × 10⁶	5.33 × 10⁶	4.98 × 10⁶	–

$T_{mc} = T_m + h_m/\rho c$ a = radius of beam, Q = thermal input power
$a = 1 \times 10^{-4}$ m, $Q = 50$ W, $t_c = \pi a^2/\kappa$, $\kappa = \lambda/\rho c$, $T_s = Q/\pi\lambda a$, $Q_{msp} = \pi\lambda T_{mc}$

and for an input power density sufficient to cause evaporation, $q = 1.10 \times 10^4$ MW/m², at $t_c = 2.66 \times 10^{-3}$ s,

$$\varphi_c = 1.04 \times 10^8 \text{ K/m}$$

Therefore, the temperature near the periphery of the projection area at the characteristic response time or at the end of the pulse width time t_c remains near room temperature, and as a result, no thermal effect occurs around the welded or pierced portion. For this reason, to pierce very small holes or to weld very narrow areas, it is preferable to use a beam of high power density, $q = 10^3$ to 10^4 MW/m², with pulses of several milliseconds.

The calculation above is based on heat conduction theory assuming that all input power is converted to heat at the surface of the workpiece. However, a high-energy electron of, say, $\sim 10^3$ keV penetrates deeply into the surface layer

Tantalum	Aluminium	Nickel	Steel	Stainless	Glass	Alumina
Ta	Al	Ni	(0.1%C)	steel (18–8)	(crown)	Al_2O_3
0.54×10^2	2.05×10^2	0.59×10^2	0.46×10^2	0.14×10^2	0.01×10^2	0.03×10^2
1.6×10^2	11×10^2	6.2×10^2	4.9×10^2	5.1×10^2	8.3×10^2	7.5×10^2
2.01×10^{-5}	6.88×10^{-5}	1.08×10^{-5}	0.12×10^{-5}	3.52×10^{-6}	0.58×10^{-6}	0.09×10^{-6}
16.6×10^3	2.70×10^3	8.8×10^3	7.85×10^3	7.93×10^3	2.4×10^3	3.99×10^3
1.57×10^{-5}	4.7×10^{-4}	0.28×10^{-4}	0.26×10^{-4}	8.93×10^{-5}	0.55×10^{-5}	0.34×10^{-5}
3.21×10^3	1.05×10^3	2.99×10^3	3.67×10^3	1.16×10^4	1.40×10^5	5.92×10^4
3270	932	1728	1673	1698	1273	2273
2773	2873	3346	–	–	–	–
–	1.13×10^9	2.68×10^9	2.85×10^9	2.30×10^9	–	–
–	2.84×10^{10}	5.81×10^{10}	–	–	–	–
–	1300	2221	2213	2224	–	–
2.19×10^6	1.58×10^6	1.01×10^6	0.84×10^6	2.66×10^5	1.34×10^4	7.11×10^3

of the workpiece and transfers its energy to the atoms. Therefore, this type of calculation is not appropriate for electron-beam processing. A detailed discussion of this will be given in the section on electron-beam processing.

Thermal data for the thermal energy-beam processing of various materials are given in Table 1.6, for an input power of 50 W and a beam of cross-sectional area about 3.14×10^{-8} m^2, i.e. of 0.1 mm diameter.

1.5 Photon and related energy-beam processing

1.5.1 The photon as an energy particle

Photon energy-beam processing may consist of a light beam or laser beam acting thermally or reactively, or of a radiation beam acting mainly reactively. Microwave energy-beam processing using thermal action is considered as a special case.

A photon is an elementary energy particle regarded as massless: the energy quantum of Planck. A light beam is made up of such photons. A photon has energy $h\nu$, where h is Planck's constant, 6.626×10^{-34} J s, and ν is the equivalent frequency (s^{-1}), since the energy particle may be simultaneously considered as an electromagnetic wave of frequency ν. The equivalent wavelength of the photon, λ (m), is given by

$$\lambda = c/\nu \tag{1.9}$$

where c is the speed of light, 2.99×10^8 m/s.

Hence, the wavelength of photons in a light beam or laser beam is of the order of 10 to 0.1 µm for photon energies in the range 0.1 to 10 eV. The wavelengths of photons in radiation such as X-rays (electron, plasma), and SOR (synchrotron orbit radiation) are of the order of 10 to 0.1 nm for photon energies of 1 to 10 keV, and 0.01 to 0.001 nm for photon energies of 0.1 to 1 MeV, respectively. Of course, the wavelength represents a statistical reaction range for the photon.

Several examples of photon-beam and related energy-beam processing, such as light-beam, laser-beam, and radiation-beam processing, are listed in Table 1.2.

The light or laser beam projected locally onto the surface of the workpiece transmits its photon energy to the outer-shell electrons of atoms. This energy is then transformed into thermal vibration energy or chemical activation energy of atoms, as shown in Fig. 1.6(a), (b), and (c).

1.5.2 Thermal processing by focused light or laser beam

In ordinary opaque substances, the energy of an irradiating photon beam is initially absorbed as thermal vibration energy in the surface atoms and is then conducted to the surrounding areas as heat.

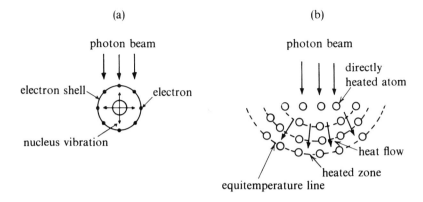
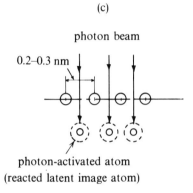

FIG. 1.6. Photon energy transfer mechanisms. (a) Photon energy to thermal energy. (b) Heat transfer to deeper areas (opaque substance). (c) Photon activation (transparent photon reactive substance)

Recently, laser beams have become widely used in this kind of energy-beam processing. A laser beam can be focused into a very fine spot of the order of 10 μm diameter. As a result, an energy beam of very high power density (10^6 MW/m^2) can be obtained in general, but in the case of a pulsed laser beam, the peak power density reaches about 10^7 MW/m^2.

There are two types of laser source available: one is a solid laser source of pulsed beams, such as YAG, ruby, etc.; and the other is a gas laser source of continuous beams, such as CO_2, etc., as shown in Fig. 1.7(a) and (b) respectively and also in Table 1.7 [10, 11]. In addition, an unstable mode of laser resonator has recently been developed to achieve a beam of higher power density, as shown in Fig. 1.8.

TABLE 1.7
Types of laser source

Laser source		Element	Optically active ion	Wavelength (μm)	
Solid	Ruby	Al_2O_3	Cr^{3+}	0.6943	Pulse, 1 J, mean max. 5 W
	YAG	$Y_3Al_5O_{12}$	Nd^{3+}	1.065	pulse, cont. 40–50 W
	Glass	Glass	Nd^{3+}	1.065	pulse, cont.
Semiconductor	GaAs	GaAs	—	~0.8	pulse, cont.
Gas	CO_2	CO_2 + He + Ne	CO_2 ion	10.63	cont. <20–30 kW pulse, 40–100 W
	Ar	Ar	Ar^+	0.488, 0.545	cont. < 5 kW
	He–Ne	He–N_2	—	0.633	cont. < 50 mW
	Excimer	(Xe, Kr, Ar) + (F, Cl, Br)	Xe^+, Kr^+, Ar^+	0.5~0.2	pulse, 1 kHz

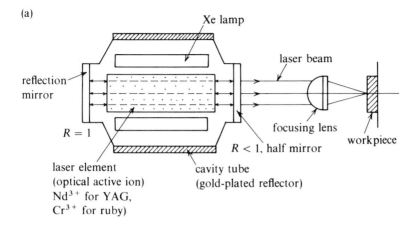

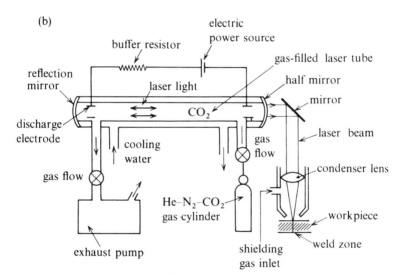

FIG. 1.7. (a) Solid laser source. (b) CO_2 gas laser source—coaxial

The high power density of the laser beam makes fast localized heating possible. Moreover, using a pulsed laser beam, small holes may be pierced in high-melting-point materials such as molybdenum and highly conductive materials such as aluminium and copper. Also, the continuous laser beam is being widely applied in manufacturing for small spot-welding or seam-welding of such materials and the cutting of sheet metal.

Other applications are as follows. Rapid heating of short duration by a pulsed laser beam, followed by rapid cooling, hardens a carbon steel surface,

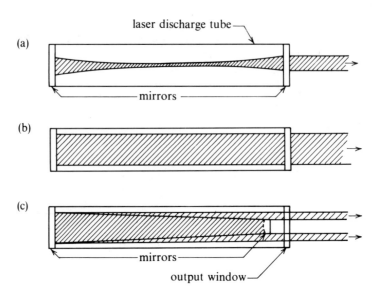

Fig. 1.8. Laser resonator modes. (a) Stable resonator—fundamental mode [TEM$_\infty$] [Gaussian] = very low power. (b) Stable resonator—multi-mode [uniform density] = 5 kW for 4-tube laser. (c) Unstable resonator—single mode = 3 kW for 4-tube laser

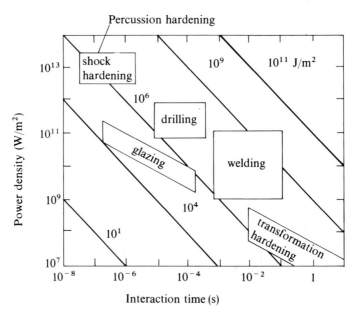

Fig. 1.9. Laser processing and irradiation (*Physics Today*, Nov. 1975)

on the principle of metallographic transformation hardening. Furthermore, with a beam of higher power density, 10^7 MW/m^2, percussion hardening or shock hardening can be performed, and a beam of 10^5 MW/m^2 is capable of glazing metal. These new applications are shown in Fig. 1.9 [12].

1.5.3 Reactive processing by light or laser beam (broad or focused)

This type of process makes use of the action of photons on photoreactive materials such as transparent photosensitive dry plates or photoreactive polymers. The mechanism of reaction is shown schematically in Fig. 1.6(c). In contrast to thermal processing, the projected photon penetrates deeply into the photosensitive material and reacts directly with a reactive atom in the path of the light beam, finally forming a centre of latent image, such as activated silver (in photography) or atoms in a polymerized or depolymerized state (in photoresist materials).

Of course, the reaction area is limited to the range of the equivalent wavelength along the light path. Hence, the argon laser and excimer laser of shorter wavelength, e.g. 0.2 μm, are preferable where high resolution is required.

Applications of this kind of process are found in the manufacture of IC wafers provided with very fine patterns of about 1 μm and of video disc masters with very fine channels and spots of about 1 μm. Details of these applications will be given later, in section 1.11.

1.5.4 Reactive processing by broad radiation beam

Beams of X-rays or SOR are also used for reactive processing. Of course, these rays consist of comparatively high-energy photons, and consequently very small equivalent wavelengths, of the order of nanometres. As a result, the radiation beam may be used successfully in the processing and micro-patterning (on the micrometre and submicrometre scales) of LSIs (large-scale integrated circuits). Using a radiation beam of a wavelength of several nanometres, the resolution can be increased to the order of 10 nm. Nevertheless, an etching process must be used after exposure; recently, however, etching using reactive-ion or low-temperature plasma beams has been increasingly used.

1.5.5 Other kinds of laser-beam processing

(i) Laser-beam-assisted chemical etching: chemical etching is promoted in regions irradiated by the laser
(ii) Laser-beam processing using holography
(iii) Laser-beam coating and plating

(iv) Laser-beam forming
(v) Laser-beam crystal growth of semiconductors
(vi) Laser-beam annealing (infra-red light beam) of ion-implanted semiconductor wafer.

These will be discussed in Chapter 2.

1.6 Electron and related energy-beam processing

Electron-beam processing may be classified into that due to thermal action, that due to reactive action, and electric-discharge processing due to thermal action, as shown in Fig. 1.1.

Electron-beam processing uses an electron beam of high acceleration voltage but low current, of the order of 10 kV and 1 mA, respectively. However, electric-discharge processing uses a pulsed spark of high current density but low acceleration voltage, e.g. about 100 A at the peak of the pulse and several volts, respectively.

Of course, an electron beam can be formed of submillimetre diameter, but in electric-discharge processing, each pulsed sparking path in the insulation oil has a diameter of several micrometres. As a result, both are beams of very high power density, of the order of 10^1 to 10^4 MW/m^2.

1.6.1 Thermal processing by focused electron beam

Schematic drawings of equipment used in electron-beam processing, and of the energy transfer mechanism from electron to atom are shown in Fig. 1.10 and Fig. 1.11, respectively.

The accelerated electrons projected onto the workpiece generally transfer their energy to electrons of the outer shells of the workpiece atoms, increasing the thermal vibration of these atoms. Consequently, the electron beam can effectively supply the heat energy necessary for thermal processing of materials. However, it is very important to recognize that electron energy is absorbed mainly in the range of the penetration depth but not at the surface of the workpiece, as is discussed later.

This processing technique was initially developed to form fine patterns on semiconductors, and fine holes and a textured surface on diamond and precious stones, because the electron has a very small diameter of 2.8×10^{-6} nm and a very small mass of 9×10^{-31} kg (as estimated from classical theory), and can be easily formed into a very fine beam of several micrometres diameter; electrons can also be highly energetized up to several hundred keV (1 keV = 1.16×10^{-16} J). However, an incident high-energy electron penetrates deeply into the subsurface layers of the workpiece through

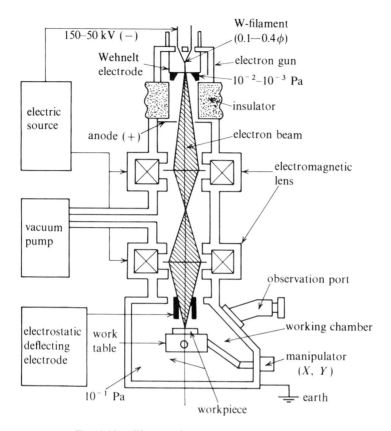

FIG. 1.10. Electron-beam processing equipment

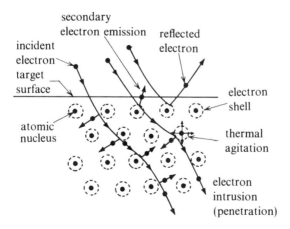

FIG. 1.11. Energy conversion mechanism of electron beam in materials

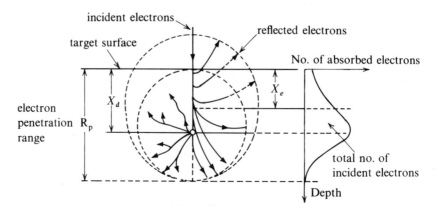

Fig. 1.12. Electron penetration range

the network of the atomic lattice, because the effective diameter of the electron is very small compared with the lattice repeat distance of about 0.2 to 0.4 nm.

Consequently, as shown in Fig. 1.12, the incident electrons dissipate their energy randomly along the penetration path and finally stop in random positions in the subsurface layers. The statistical electron penetration depth, R_p(m), in which the great majority of penetrating electrons are absorbed, is obtained experimentally and theoretically as

$$R_p = 2.2 \times 10^{-11} V^2/\rho \quad (m) \tag{1.10}$$

where V is the acceleration voltage (V), and ρ is the mass density of the workpiece (kg/m^3). For example, the penetration depth R_p in steel for electrons of energy 50 keV is about 7 μm, and in aluminium, about 10 μm.

Moreover, the number of electrons absorbed at the workpiece surface is very small, as shown in Fig. 1.12. The depth of maximum absorption rate is X_e; however, a large proportion of incident electrons are absorbed around depth X_d [13]. Consequently, the surface atoms of the workpiece cannot be heated directly by the projected electrons, but are heated by conduction from the subsurface layer. Therefore, thermal processes such as melting and vaporization cannot be limited to a very thin layer only a few micrometres in depth, and thermal machining or welding by electron beam cannot be expected to be successful for very fine processing at the micrometre scale, even though the electron beam can be formed to a diameter less than this. For this reason, practical welding and stock removal processes employ electron beams only for larger-scale operations. The processing mechanism is explained in terms of bubble splashing, in Fig. 1.13.

The most important application of electron-beam processing is deep-penetration welding. This process, making use of a vapour bubble, is

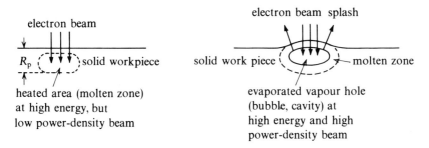

FIG. 1.13. Models showing electron-beam melting and stock removal by splashing

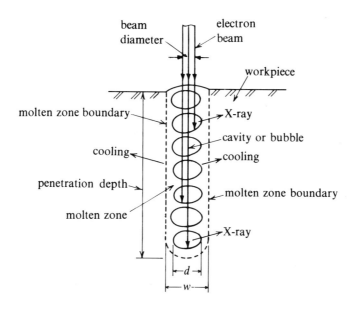

FIG. 1.14. Electron beam deep penetration model

explained by the deep-penetration model in Fig. 1.14 [14]. Typical data for deep-penetration welding using a stationary electron beam are given in Fig. 1.15. The data show that for deep-penetration welding of steel, a beam of high power density, >10 MW/m^2, is necessary.

A specially designed zoom electron-beam processor for deep welding is shown in Fig. 1.16. Practical welding techniques, electron-beam annealing, etc. will be discussed in Chapter 3.

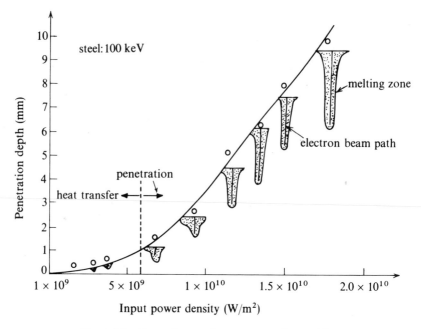

Fig. 1.15. Deep electron-beam penetration welding

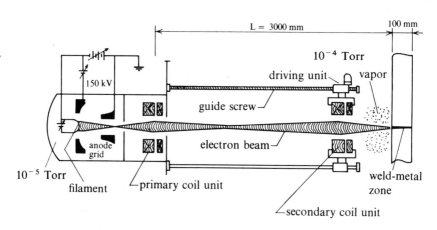

Fig. 1.16. Zoom electron gun for welding
(after Hitachi Corporation, Japan)

1.6.2 Reactive processing by focused electron beam

Recently, in order to make extremely fine patterns of a few micrometres for ICs (integrated circuits) on semiconductors, increasing use has been made of this kind of electron-beam processing. Electron-beam exposure or electron-beam lithography, is shown in Fig. 1.17.

This process is based on the activation of electron-sensitive materials such as polymers by an electron beam. Atoms of the electron-sensitive material are activated along the path of an incident penetrating electron. The outer-shell electrons of the atom are activated by the incident electron, and polymerization or depolymerization occurs, as in photon activation. This process has been used to achieve very high dimensional resolution, because no thermal

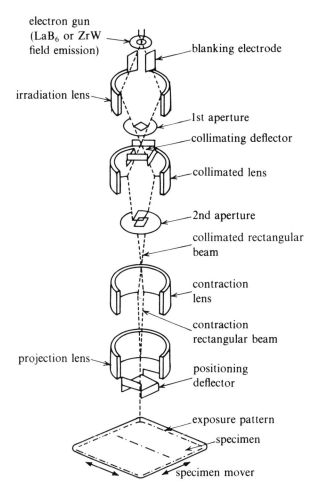

FIG. 1.17. Electron-beam lithography. Variable-area electron beam exposure equipment JBX 6A (vector scanning type). There is also a raster scanning type, which deflects its electron beam horizontally.

energy diffusion occurs and there is only slight secondary electron scattering. Electron image projection and other electron-beam processes, are described in Chapter 3.

1.6.3 Electric discharge beam processing (tool-guided beam)

Electric discharge beam processing is shown in Fig. 1.1 as one type of electron energy-beam processing. An energy beam of high power density, about 10^4 to 10^5 MW/m^2, is obtained by electric sparking through a conductive plasma path, several micrometres in diameter, in the insulating oil between the metallic workpiece and the electrode tool. In this way the discharge sparking point or electron bombardment point on the workpiece surface is heated

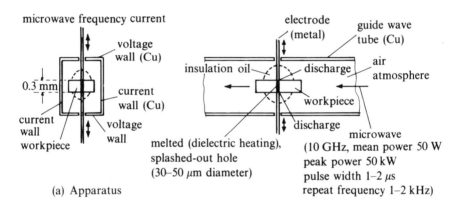

(a) Apparatus

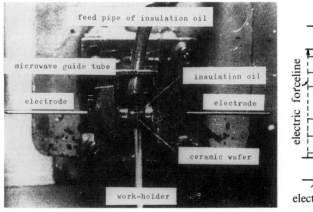

(b) Apparatus.

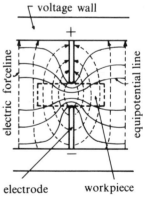

Electric field near electrode

FIG. 1.18.

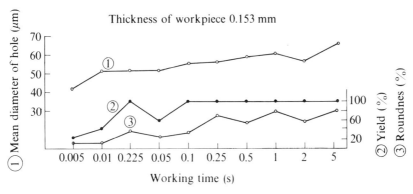

(c) Relation between the shape of pierced hole and working time.

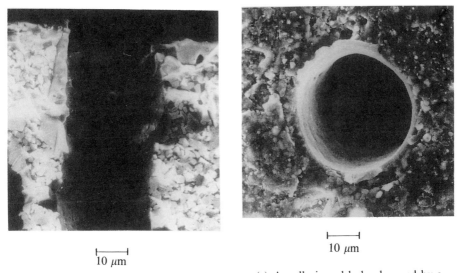

(d) Inner surface of the pierced hole.

(e) A well pierced hole observed by a scanning electron microscope.

FIG. 1.18. Microwave current beam piercing of alumina ceramic wafer by dielectric heating. (**a**) and (**b**) Apparatus (10 GHz, mean power 50 W, peak power 50 kW, pulse width 1–2 μs, repeat frequency 1–2 kHz). (**c**) Relation between shape of pierced hole and working time. (**d**) Inner surface of picrced hole. (**e**) A well pierced hole observed by a scanning electron microscope

rapidly and locally to vaporization temperature and material is removed in the form of a melted splash.

This process, developed since 1940, has already become a conventional machining process. It is called electric discharge machining (EDM) and many books have been published describing the process. However, here we shall draw attention to electric spark machining of dielectric insulators, such as diamond (direct discharge) and ceramics (dielectric loss), using microwaves of 10 GHz with a pulsed beam of high power density, as shown in Fig. 1.18 (mainly for piercing of ceramic wafers). For details of the process, refer to the author's paper [15].

1.7 Ion and related energy-beam processing

This is classified into dynamic and reactive processing.

1.7.1 Dynamic processing by focused or broad ion beam

The ion beam consists of highly energetic ions of an inert gas, such as argon, krypton, or xenon, which can be easily accelerated and controlled by electrical means. Removal by dynamic processing is carried out by knocking atoms out of the surface of the workpiece by electro-elastic collision and recoil action between the projected ion and the target workpiece atom. This is called ion sputtering.

(1) Fundamental ion-sputter phenomenon

As shown in Fig. 1.19, electrically accelerated inert gas ions, such as Ar ions of average energy 10 keV (corresponding to a speed of about 200 km/s) are unidirectionally oriented and projected on to the workpiece surface in a high vacuum of 10^{-6} Torr (1.3×10^{-4} Pa).

In contrast to electron-beam processing, most of the projected ions interfere with surface atoms of the workpiece, because the mean diameter of the Ar ion is about 0.1 nm and the mean lattice distance of the workpiece atoms is about 0.3 nm. Consequently, the projected energetic ions frequently collide with the nuclei of atoms of the workpiece, and recoil and knock out, or sputter out, atoms from the workpiece surface. Hence, the process is performed atom by atom, and is called ion sputter etching or ion sputter machining.

Furthermore, the sputtered atom has an energy of 100 eV, greater than that of an ordinary evaporated atom. By virtue of this, ion sputter deposition may be performed as follows: another workpiece is set up opposite the initial target; and the sputtered atom collides with it and adheres more firmly than it would as a result of ordinary vapour deposition.

The penetration depth of an impinging Ar ion of 10 keV is estimated at several nanometres or of the order of ten atomic layers of depth, as shown on the left of Fig. 1.19. However, for higher-energy Ar ions of about 100 keV, the penetration depth is greater, because an ion of higher energy can pass through the atomic lattice network more freely, gradually losing its energy. The deeply penetrating ion with reduced energy may then collide with an atom of the

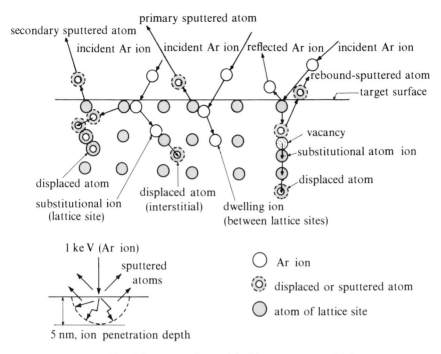

FIG. 1.19. Diagrammatic model of ion sputter machining

workpiece and knock it from its lattice site. However, it is easily seen that an atom knocked out at such a deep position cannot be sputtered from the workpiece surface and remains as a substitutional or interstitial atom. Consequently, the process using higher-energy ions is not applied in ordinary ion sputter etching, but it is widely used in ion implantation processes to inject impurities into semiconductors. Typical equipment used in ion implantation is shown in Fig. 1.20.

(2) Ion sputter machining characteristics and equipment

The effective acceleration voltage, or ion energy, for this process appears to be about 20 to 30 keV for Ar and Kr, and 40 to 50 keV for Xe, on SiO_2, as shown in Fig. 1.21.

The relationships of the sputtering yield (i.e. ratio of number of sputtered atoms to that of projected ions) and the rate of machining (by depth) to the ion incidence angle are shown in Fig. 1.22. The sputtering yield is greatest at an incidence angle of about 80°, but the machining rate is greatest at about 60° (but compare Figs. 4.7 and 4.8 in Chapter 4, and see Section 4.2.1(2) in that chapter).

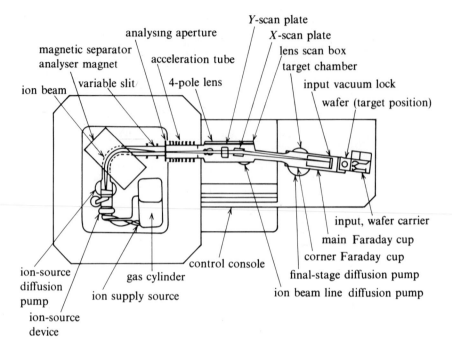

Fig. 1.20. Ion implantation equipment

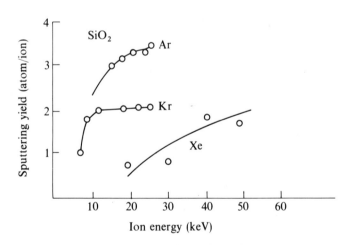

Fig. 1.21. Ion sputtering ratio vs. ion energy

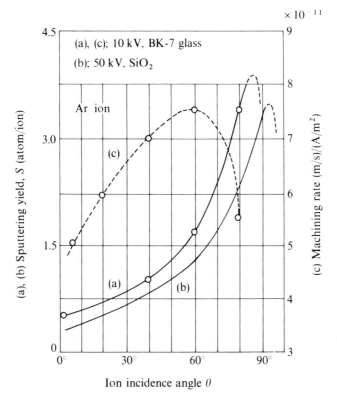

FIG. 1.22. Ion sputter machining characteristics

These machining characteristics are very important and have been made use of in the development of various applications of ion sputter etching, such as the resharpening of diamond point tools, and the production of diffraction gratings and aspheric lenses [16–20].

As regards equipment used in ion sputter machining, there are three types of ion-beam source: the focused ion-beam type as shown in Fig. 1.23; the high-frequency plasma type as shown in Fig. 1.24; and the ion-shower type as shown in Fig. 1.25. Detailed explanation will be given in Chapter 4.

(3) Ion sputter processing as atomic-scale machining

In this type of process, machining or etching is performed atom by atom, so the machining resolution is very fine, compared with thermal energy-beam processing. However, unlike a solid tool, the ion beam cannot directly generate a desired geometric form, because the projection point of the ion beam cannot be controlled precisely enough to provide sufficient working

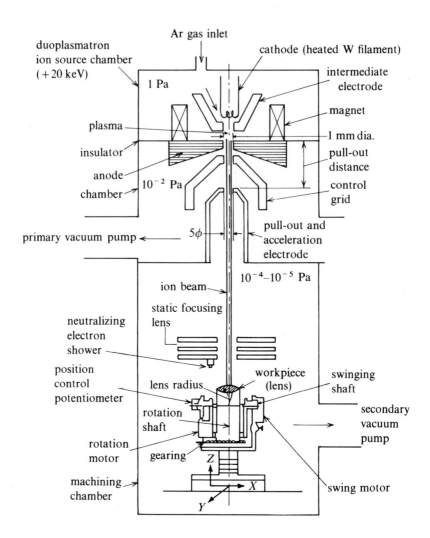
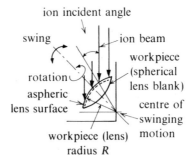

FIG. 1.23. Ion-beam source equipment: focused ion-beam type for aspheric lens-making

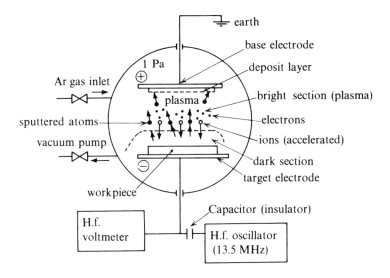

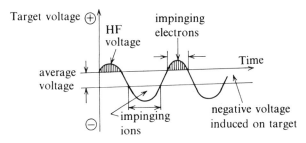

FIG. 1.24. Ion-beam source equipment: high-frequency plasma type

accuracy, owing to some randomness in an ion beam based on a gaseous ion source, and to the lack of a position feedback system. Also, machining depth is determined only by machining time.

Therefore, two-dimensional machining of high accuracy and fineness must be carried out using masks of high accuracy; three-dimensional machining is performed by a uniform and stable ion beam, with constant energy controlled by a magnetic ion separator, as shown in Fig. 1.20.

Micro ion beams of high brightness of the needle and capillary types have recently been developed, using readily fusible substances such as Ga. They are called liquid-metal ion sources (LMIS) and electrohydrodynamic ion sources (EHDIS). As a result, an ion beam of about 1 to 100 μm diameter is currently being applied for micro-patterning without a mask in wafer processing of very high-density integrated circuits.

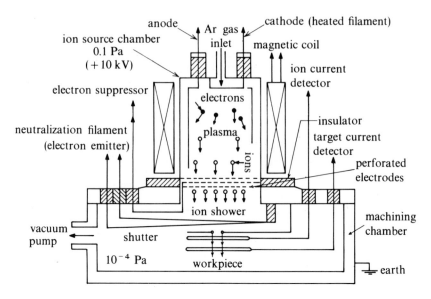

FIG. 1.25. Ion-beam source equipment: ion-shower type. (D.c. discharge; microwave resonator; 2.45 GHz)

Nevertheless, in the near future, in-process dimensional measurement and control methods with resolution, sensitivity, and accuracy of the order of 0.1 nm or atomic size, and high-resolution, rapid-response servo systems will be developed. This kind of ion beam is becoming a most useful tool in 'nanotechnology', which is the name given to manufacturing processes for very fine and accurate components, with accuracies of the order of nanometres.

Applications of dynamic ion-beam processing such as ion mixing, ion rubbing, ion deposition, ion plating, ion nitriding, and ion-assisted chemical etching, etc. will be discussed in Chapter 4.

(4) Power density in dynamic ion-beam processing

Processing by means of dynamic ion action is on the atomic scale, with apparently very low power density, as shown in Table 1.5, but microscopically the process is of very high power density, as explained by the following reasoning.

In this type of process, the duration of electro-elastic collision between incident ion and target atom is estimated to be about 10^{-12} s (this estimate is based on the approach speed of the incident ion—about 200 km/s). The kinetic energy of the target atom is increased as a result of the momentum change in this collision. Moreover, the kinetic energy necessary for the target atom to be knocked out or sputtered is assumed to be 10^{-18} J/atom, estimated from the atomic bonding energy. Therefore the input power for knocking out the target atom can be estimated as $10^{-18}/10^{-12} = 10^{-6}$ W/atom. The apparent input power density becomes 10^8 MW/m², because the equivalent atomic collision area corresponding to a van der Waals atomic radius of 0.2 nm is 4×10^{-20} m²/atom.

Consequently, this kind of dynamic ion-beam processing method can be said to be a process of very high power density. That is to say, it is an atomic-scale process, and even though the macroscopic input power density is small, the microscopic or atomic-scale input power density is extremely large. As shown in Table 1.5, the mean macroscopic input power density P_m is about 10^{-2} to 1 MW/m², owing to limitations imposed by technical problems, but microscopically, the input power density, P_a, becomes about 10^6 to 10^7 MW/m².

On the other hand, the macroscopic machining speed of this process v (m/s) can be calculated, using values of mean input power density P_m (W/m²) and threshold specific processing energy density for the atomic lattice range, $\delta = 10^6$ to 10^5 MJ/m³, as shown in Table 1.3. For example:

$$v = P_m/\delta = 10^6/10^{12} = 10^{-6} \text{ m/s} = 1 \text{ } \mu\text{m/s}$$

Accordingly, if a new method is developed to supply input power at a higher macroscopic density, the machining speed (by depth) will be greatly improved.

1.7.2 Reactive processing by focused broad ion beam

The atomic-scale removal process making use of chemically reactive ions, derived from CCl_4 for aluminium or CF_4 for silicon, is performed at considerably higher speed than in ordinary chemical processing, because the process is accelerated by the activation energy supplied by the projected ion. Of course, this means that the macroscopic machining speed, or statistical rate of processing, is increased. Detailed discussion is presented in Chapter 4.

Etching may also be performed by making use of the nascent state of reactive ions derived from CCl_4 and CF_4 in a cold plasma produced in a high-frequency source or an ion-shower source, as shown in Fig. 1.24 and Fig. 1.25 respectively. These processes are widely used for etching wafers in the semiconductor industry.

1.8 Atom or molecule energy-beam processing

This kind of energy beam consists of neutral atoms or molecules, which are obtained by thermal evaporation caused by electron-beam heating in a high vacuum of 10^{-9} to 10^{-10} Torr (about $10^{-7} \sim 10^{-8}$ Pa). The kinetic energy of atoms or molecules in this process is very small (about 1 eV), in contrast to that of sputtered atoms. Direct deposition as a result of evaporation in a lower vacuum is used for thin-film making, but this process is more widely used for epitaxial crystal growth on semiconductor wafers, by virtue of its fine controllability of thickness and composition. A synthetic process for 'super

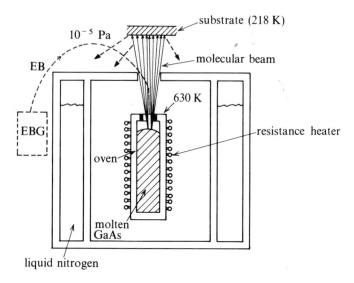

Fig. 1.26. Molecular-beam epitaxy (MBE)

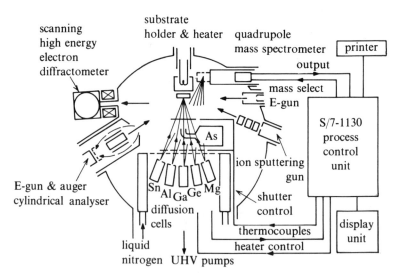

Fig. 1.27. Crystal growth due to molecular beam controlled by digital computer

lattice substance' for microelectronic components has also been developed using this system. Typical equipment is shown in Figs. 1.26 and 1.27. This is called cold molecular-beam processing. Hot flame welding and hot metal spraying may be designated as hot molecular-beam processing. However, a water jet process should be treated as a mechanical process.

1.9 Plasma energy-beam processing [21, 22]

Plasma is defined as an electrically conductive state of gas, where approximately equal numbers of electrons and ions are simultaneously present. Generally speaking, at atmospheric pressure, plasma appears in arc discharge gas at temperatures of 10 000 to 20 000 K, as shown in Table 1.8. That is, atoms of inert gas and H_2, O_2 etc. initially disintegrate into ions and electrons, under bombardment by free electrons accelerated in the gas discharge.

TABLE 1.8
Types of high-temperature source

	Temperature (K)	Speed (m/s)	Heat generation rate (W/m^2)	
Bunsen burner	1900	6	4.2×10^7	CH_4 etc.
Oxygen–acetylene	3300	120	3.2×10^8	O_2, C_2H_2
Shielded arc (TIG, MIG, CO_2)	8000 ~ 10 000	600 ~ 1000	5.1×10^9	He
Plasma arc beam	10 000 ~ 20 000	1000 ~ 3000	8.4×10^9	He, Ar, N_2
Plasma jet beam	20 000 ~ 50 000	3000 ~ 5000	1.6×10^{10}	He, Ar, N_2

However, in the plasma generated in an arc discharge, some ions and electrons recombine and the resulting energy is transformed into thermal vibration energy of the reunited atom, or is released as a photon. This state of disintegration and recombination is characteristic of plasma..

Plasma-beam sources of the arc and jet types are shown in Fig. 1.28(a) and (b). Plasma-arc beam processing is performed by plasma produced by direct arc discharge against the workpiece, and plasma-jet beam processing by high-speed plasma blown from a plasma gun. Figure 1.29 shows a modified plasma gun which is very convenient in controlling plasma power. Ordinary electric arc welding should also be considered as plasma processing.

For detailed explanations of plasma-beam processing, the reader should refer to specialist books.

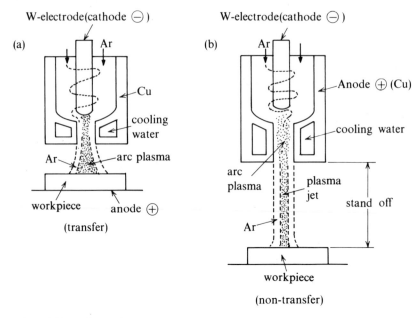

FIG. 1.28. Plasma beam sources. (a) Plasma arc beam. (b) Plasma jet beam

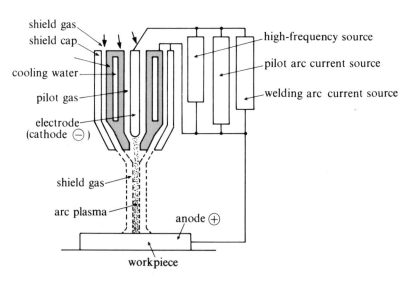

FIG. 1.29. Modified plasma arc beam source
(Nippon Minibearing Co.)

1.10 Chemical and electrochemical reactant energy-beam processing (broad beam and tool-guided beam)

These processes are performed by a beam of chemical or electrochemical reactant in the gas or liquid phase. In removal processes, the reaction products, which are workpiece atoms combined with the reactants, should be removed from the workpiece surface by a forced flow of reactant gas or liquid. It is necessary to apply forced circulation or thermal convection to promote the chemical reaction, because reaction products on the workpiece inhibit subsequent reaction.

Various kinds of chemical and electrochemical reactant energy-beam processing for removal, joining, forming, and surface treatment are listed in Table 1.2.

1.10.1 Fundamentals of reactive energy-beam processing

As shown in Fig. 1.30, the internal free energy is related to the state of the reaction. The elementary reaction process is governed by the molar activation energy Q_a (J/mol) and initial and final states of the molar Gibbs energy level, G_i, and G_f respectively (J/mol). The difference between both free energy levels,

$$\Delta G = G_i - G_f$$

is called the driving potential when $\Delta G > 0$, and the resisting potential when

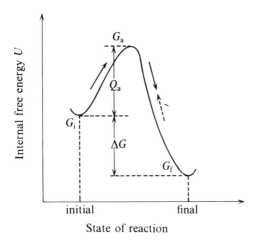

FIG. 1.30. Activation energy. G_a = Gibbs energy of activated state; G_i = Gibbs energy of initial state; G_f = Gibbs energy of final state; Q_a = activation energy; $G = G_i - G_f$

$\Delta G < 0$. However, ΔG governs only the direction of the reaction, not the reaction rate.

The reaction rate in chemical processing is determined statistically by the number of activated atoms whose molar internal energy U is increased by the activation energy Q_a above the Gibbs free energy G_i or G_f, as in the well-known reaction rate theory of Arrhenius.

Let n be the number of reacting atoms per mole, and Q_a be the activation energy; then n is determined by the energy distribution theory of Maxwell and Boltzmann, as shown in Fig. 1.31. That is,

$$n \propto \exp[-(Q_a - \bar{E})/kT] \tag{1.11}$$

where $\bar{E}$ (J/mol) is the molar mean thermal energy at the reaction temperature T (K), and k is the Boltzmann constant, 1.38×10^{-23} J/(K mol). Therefore, the reaction rate r (s^{-1}) is determined by the ratio n/N, where N is the total number of reacting atoms per mole. Then

$$r \propto (n/N) = A \exp[-(Q_a - \bar{E})/kT] \tag{1.12}$$

where $A = f(c)pv \exp(S_a/k)$ is the reaction rate coefficient, c is the concentration of reactive atoms or molecules, S_a is the molar entropy of the reacting atoms, v is the frequency of atomic thermal vibration (s^{-1}), and p is the workpiece form coefficient.

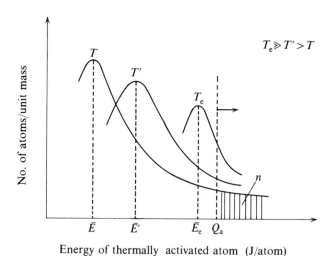

Energy of thermally activated atom (J/atom)

FIG. 1.31. Distribution of number of thermally activated atoms. T, T' = temperature; T_e = evaporation temperature; $\bar{E}$ = mean energy of thermally activated atoms; n = number of atoms having energy greater than Q_a (activation energy)

In general, $Q_a \gg \bar{E}$, so the reaction rate r is expressed as follows:

$$r = A \exp(-Q_a/kT) \quad (1.13)$$

Now let us consider the reaction from G_i to G_f. The reaction rate r_1 in the direction of the driving potential $\Delta G > 0$, is given by

$$r_1 = A_1 \exp(-Q_a/kT) \quad (1.14)$$

and the reaction rate r_2 in the direction of the resisting potential $\Delta G < 0$ by

$$r_2 = A_2 \exp[-(Q_a + \Delta G)/kT] \quad (1.15)$$

as shown in Fig. 1.30. Then the resultant reaction rate r_t at the same temperature and at the same reaction rate coefficient, $A_1 = A_2 = A$, is obtained as follows:

$$r_t = r_1 - r_2 = A \exp(-Q_a/kT)[1 - \exp(-\Delta G/kT)] \quad (1.16)$$

Accordingly, when $\Delta G > 0$, the resultant reaction rate r_t becomes positive and true reactive processing takes place, but when $\Delta G < 0$, r_t becomes negative and no actual reactive processing occurs.

In a recent development, therefore, a direct activation system involving a laser beam or electron beam is used to excite a macroscopic reaction or to increase the reaction rate in a localized area; such a system has been applied particularly to the wafer processing of semiconductors.

This kind of chemical-beam processing is of atomic scale for removal and consolidation, so it is widely used for chemical polishing, chemical plating, etc. However, recently, in the micro-fabrication of LSIs, the process has been found especially effective when used for photo-etching and other chemical surface treatments.

Electrochemical processing can be treated by a similar reaction theory, but essentially the process is governed mainly by the electrochemical reaction due to the ionization of the reacting liquid; this type of processing technique is used mainly for removal or consolidation of conducting materials.

1.10.2 Input power density and processing speed of chemical and electrochemical processes

As shown in Table 1.6, macroscopic input power densities P_m for these processes appear to be very small and are estimated as 10^{-2} to 10^{-1} MW/m². The macroscopic processing speed (by depth), v (m/s), obtained experimentally, is about 10^{-1} to 1 μm/s. The power density P_m is determined from the equation $P_m = v\delta$, where δ is assumed to be 10^5 MJ/m³.

Assuming the net reaction time to be 10^{-10} to 10^{-11} s, estimated from the atomic thermal vibration frequency, then the microscopic input power is evaluated as about 10^{-8} to 10^{-7} W/atom, calculated from the value of the lattice bonding energy of 10^{-18} J/atom. The microscopic atomic input power density is estimated as about 10^5 to 10^6 MW/m^2, based on an apparent atomic working area of $(2\times 10^{-10})^2$ m^2. This value is considerably smaller than that in ion-beam processing.

In deposition or consolidation by chemical or electrochemical processing, this level of processing power density is necessary, and in order to improve the processing rate, the problem of how to increase the input power density in practice must be considered a priority.

1.11 Micro-fabrication techniques using energy-beam processing of materials [1, 23]

Recently, many microfine structures have been widely used in electronic devices. This section gives two examples of microfabrication processes for MOS (metal oxide semiconductors): the IC transistor and the video disc. As shown in the following paragraphs, the main steps in the fabrication of these products involve various types of energy-beam processing.

1.11.1 *Fabrication of an MOS IC transistor*

Cross-sections of the functional pattern of an MOS transistor as a circuit element of an IC (integrated circuit) are shown in Fig. 1.32(a) and (b), where (a) represents a very small MOS transistor in a super LSI (large scale IC) and (b) is an MOS transistor from a 16 kbit CC RAM (random access memory) IC. As can be seen, the smallest dimension of the pattern in the MOS transistor is about 0.24 μm in the LSI, but in an ordinary MOS transistor it is about 3 μm.

The micropattern fabrication steps for transistors on IC wafers are shown in Fig. 1.33, where (a) is a plan view of a silicon wafer sliced from a silicon crystal rod and mechanically and chemically polished, and (b) is an internal view of one chip, having several hundreds or thousands of transistors, as shown in (c); (d) shows the wafer patterning process, consisting of several steps, such as exposure, development, stripping, doping, and annealing; and finally (e) shows the preparation of a silicon wafer for the exposure process.

The most important wafer-patterning process is the exposure of the photoresist (1), spread on an SiO$_2$ layer of the silicon wafer, using a pattern mask and ultra-violet radiation. Processing is performed by the mask alignment equipment. This is photon-beam reactive processing, but recently, electron-beam and X-ray exposure have been applied.

After this exposure process (1) (Fig. 1.33(d)), the exposed part is developed by selective wet etching and thus the photoresist layer is provided with patterned holes. The next step is the etching of the SiO$_2$ film layer under the patterned photoresist layer (3). This is performed mainly by dry processes

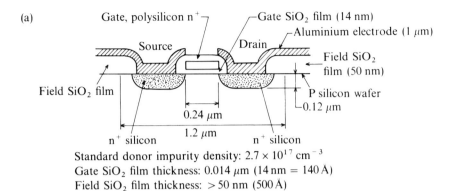

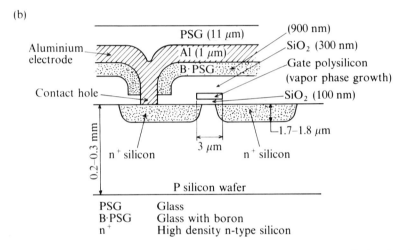

FIG. 1.32. (a) Sectional view of minimum dimensional MOS transistor. (b) Sectional view of 16 kb CCRAM MOS transistor

such as the use of a reactive ion beam or O_2 plasma beam. The subsequent stripping of the photoresist layer is effected by the same O_2 plasma beam (4).

By the processes of doping (5), using thermal diffusion or ion-implantation processing, and annealing (6), using infra-red radiation or a laser beam, the functional patterned semiconductor chips are produced on the silicon wafer.

In the deposition of functional layers on an SiO_2 film, the several processes of vapour deposition, ion-beam deposition, and molecular-beam epitaxial growth are performed as shown by (5) and (6) in Fig. 1.33(d).

In practice, repeated steps of wafer patterning become necessary, because of the complexity of the patterns, as seen in the cross-section of the MOS IC.

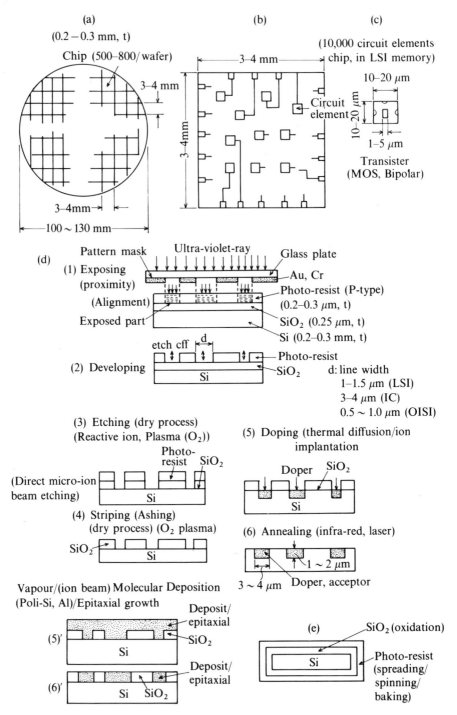

FIG. 1.33. IC wafer pattern process: (a) silicon wafer; (b) chip; (c) circuit element; (d) wafer patterning process (sectional view); (e) preprocessing for exposure

Accordingly, it becomes most important to align precisely the succeeding pattern of the mask with the pattern already processed on the wafer.

In Fig. 1.34, pattern-mask lithography and pattern-exposure systems are shown. The latest most advanced exposure system for LSIs is considered to be the direct step and repeater (DSR), i.e. chip-by-chip aligner, which has a working resolution of linewidth (WRES) 1 μm, and repeated positioning accuracy (RPA) 1 to 0.05 μm. However, in an ordinary IC having a smallest critical dimension of several μm, wafer-by-wafer aligners, such as contact, proximity, and projection aligners, which have an RPA of 0.2 μm, are used.

For making pattern masks for ICs, a pattern generator, provided with a numerical controller, employing no original pattern, is currently used, but soon there will be increased use of many advanced direct systems of very high accuracy, such as electron-beam, excimer-laser-beam, and micro-ion-beam lithography systems.

Furthermore, for the finest pattern exposure systems requiring submicrometre linewidths, X-ray (SOR, synchrotron orbit radiation) beam exposure systems of submicrometre resolution power will be used.

The yield on the IC production line is a most important concern for IC manufacturers. At present the maximum potential yield is not achieved, because wafer processing as described above causes several inherently unavoidable errors in the patterns on the wafer at each step, owing to random etching rates, random deformations in high-temperature diffusion processing, and integrated alignment errors, etc. The actual yield achieved on ordinary IC production lines is given as about 60 to 70 per cent, and for LSIs, about 20 to 30 per cent. The yield may be greatly improved by recently developed processing systems and by inspection and control systems in the dustless clean room.

1.11.2 Fabrication of video discs

There are two types of video disc available; one is an electric capacitance type and the other an optical type, as shown in Fig. 1.35(a) and (b), respectively. Both types are provided with nearly the same area for information signals on the disc plate, which has a channel 0.6 to 1 μm in breadth, 0.1 to 0.3 μm in depth and of varying length. However, the information pick-up systems differ from one another. The capacitance type picks up the information signals by a change in electric capacitance between the plated area on the disc and the electrode of the pick-up needle, which consists of diamond plated with high-melting-point metal, as shown in (a). The optical type, on the other hand, picks up the information signals by a change in the reflection of a projected laser beam from the surface of the disc, as shown in (b); the changes are monitored by electronic and optical sensors.

The most important process in disc fabrication is the microfine coining or

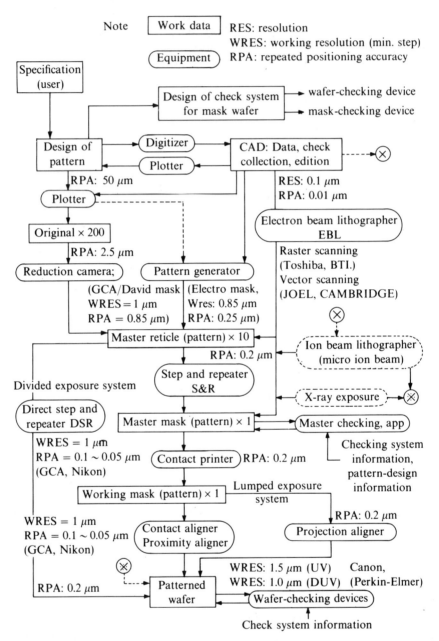

FIG. 1.34. Pattern Mask Lithography and accuracies of related apparatus

stamping process, using a very accurate stamper with a very fine pattern. The disc fabrication system should be suitable for large-scale production, to allow a reasonable price, so it is important to make the stamper cheaply. In practice, an electroplating process is used for stamper production, as shown in Fig. 1.35(c) and (d).

However, recently, high-speed injection moulding systems have been increasingly applied in disc-making, especially for audio discs.

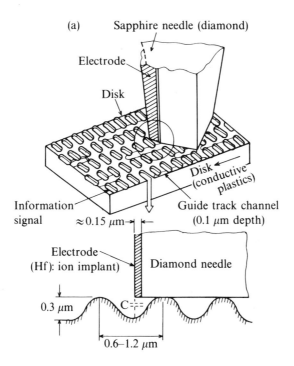

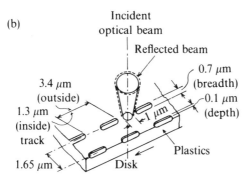

FIG. 1.35 (a and b).

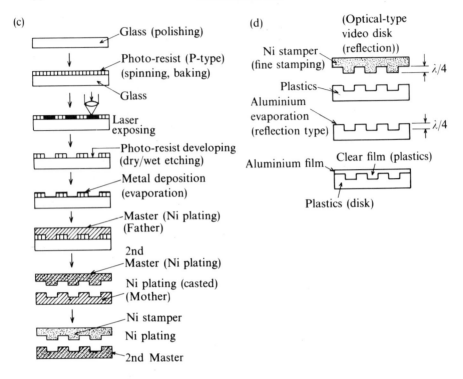

Fig. 1.35. Video disk, optical disk, and stamper; (a) capacity-type video disk; (b) optical-type video disk; (c) stamper-making process; (d) disk stamping

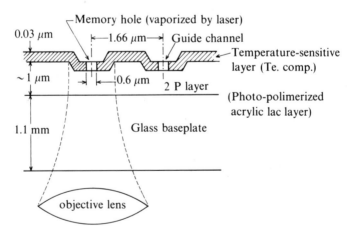

Fig. 1.36. Optical disk DRAW

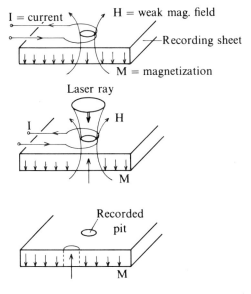

FIG. 1.37. Optical disk DRAW and erase (opto-magnetic system)

In contrast to the above-mentioned video disc and acoustic disc, the optical disc capable of DRAW (direct read after write) has recently been developed. The signal size of the disc is about 1 μm, the same size as that of the video disc, as shown in Fig. 1.36. However, several kinds of erasable optical discs capable of DRAW are being developed. One of these is an optical–magnetic system as shown in Fig. 1.37 and another is an optical phase-change system. For details of these discs, refer to special reports.

1.12 Summary

The basic characteristics of energy-beam processing of materials have been analysed in terms of the concept of processing energy.

The threshold specific processing energy δ is estimated at about 10^3 to 10^2 MJ/m^3 for the processing unit size $\varepsilon - 10^{-1}$ to 10 μm, or dislocation or microcrack range, which is the smallest unit size which can be processed mechanically.

The threshold specific processing energy δ for the processing unit size of atomic lattice range is larger, about 10^6 to 10^4 MJ/m^3, in atomic-scale energy-beam processing, as shown in Table 1.3. To obtain this high level of specific processing energy, the energy beam should be of high power density, i.e. 10^3 to 10^4 MW/m^2 for a continuous beam, or 10^4 to 10^6 MW/m^2 for a pulsed beam. This is found in electron-beam and laser-beam processing.

In ion-beam processing, and chemical or electrochemical processing, the macroscopic input power density P_m is very low (10^{-2} to 10^{-1} MW/m^2), but at the microscopic or atomic level, the peak input power density P_p becomes very large (10^5 to 10^6 MW/m^2 for chemical or thermal processes, and 10^6 to 10^7 MW/m^2 for ion-beam processes, as shown in Table 1.5).

Details of photon-beam, electron-beam, molecular-beam, and ion-beam processing are given in the later chapters of this book, but for details of plasma-beam and chemical or electrochemical processing the reader should refer to specialist books. For an outline of recent developments in energy-beam processing of materials see the author's papers [1, 23].

Applications of energy beams in manufacturing processes in the semiconductor industry in particular are expanding rapidly; such beams are very suitable for ultrafine and ultra-precision processing of integrated circuits (IC). Processing is performed basically atom by atom, and therefore the resolution is extremely high; of the order of nanometres. Atomic-scale deposition processes are also very promising for the semiconductor and precision materials industries.

References

[1] Taniguchi, N., 'Research on, and development of energy-beam processing of materials in Japan', *Bull. Japan. Soc. of Prec. Eng.* **18** (1984), 2.

[2] Taniguchi, N., 'Analysis of various materials working, based on the concept of working energy', *Scientific papers of the Inst. of Phys. and Chem. Research (Japan)* **61** (1967), 3.

[3] Taniguchi, N., 'On the basic concept of nano-technology', *Proc. of ICPE (Int. Conf. on Prod. Eng.), Tokyo* (1974).

[4] Taniguchi, N., 'Current status in, and future trends of ultra-precision machining and ultra-fine materials processing', *Annals of the CIRP (Int. Inst. for Prod. Eng. research)* **32** (1983), 2.

[5] McClinton and Irwin, 'Fracture toughness testing' (ASTM STP No. 381).

[6] Beck, W. R. et al., *Trans ASME*, **74** (1952).

[7] Plendle, J. N., *Phys. Review*, **123** (1961), 4 and **125** (1962), 3.

[8] Carslaw, H. S. and Jaeger, J. C., *Conduction of heat in solids*, 2nd edn., (Oxford University Press, 1959).

[9] Taniguchi, N. and Maezawa, S., 'Pulsed beam temperature analysis of electron-beam machining', *Proc. of 5th Annual Electron Beam Symp.*, (Boston, 1963).

[10] Raytheon Co., *Laser and materials processing outline information* (1971), 1.

[11] Cohen, M. I. and Epperson, J. P., *Electron beam and laser beam technology* (Academic Press, N Y 1968).

[12] Muraoka, M. et al., *O pulse E.* (1983), 8.

[13] Kanaya, K. and Okayama, S., *Proc. 7th Int. Cong. Electron Beam Microscopy, Grenoble* **II** (1970), 159–60.

[14] Miyazaki, T. and Taniguchi, N., *Proc. of Electron Beam Science and Technology*, 5th Int. Conf. (1972).
[15] Taniguchi, N. et al., 'On the electromagnetic microwave machining of ceramic wafers', *Annals of CIRP* (1972).
[16] Wilson, R. G. and Brewer, G. R. *Ion beams*, (John Wiley & Sons, Inc. 1973).
[17] Taniguchi, N. and Miyazaki, T., *Annals of CIRP* **23** (1974).
[18] Taniguchi, N., Yoshimoto, S., and Miyamoto, I., *Annals of CIRP* **24** (1975).
[19] Miyamoto, I. and Taniguchi, N., *Proc. 3rd Int. Conf. on Prod. Eng., Kyoto* (1977).
[20] Taniguchi, N., *Annals of CIRP* **32** (1982), 2.
[21] *Nontraditional machining* (Machinability Data Center, Metcut Research Associate Inc.)
[22] *Tool and Manufacturing Engineers Handbook* (Society of Manuf. Engineers, 1983).
[23] Taniguchi, N., 'Atomic bit machining by energy-beam processing', *Precision engineering* **7**, No. 3 (July 1983).

2
PHOTON-BEAM PROCESSING

2.1 Introduction

As mentioned in Chapter 1, photon-beam processing using a laser is called laser-beam processing. In 1960, Maiman, of the Hughes Corporation, succeeded in the first demonstration of oscillation of pulsed 694.3 nm light, using a synthetic ruby rod 6 mm in diameter and 45 mm in length [1]. Since then, many kinds of laser have been developed: in 1961, an He–Ne laser by Javan of Bell [2]; an Nd–glass laser by Snitzer of American Opticals [3]; in 1962, a semiconductor laser [4]; in 1964, an Ar ion laser [5], a Nd–YAG laser [6], a CO_2 laser [7]; and in 1966, a liquid laser [8].

Recently the laser has been increasingly used in fields such as materials processing, measurement, communication, and medical science. Material processing in particular has, since the early stages, the most widespread application. The first practical application was in drilling a diamond (for a wiredrawing die) with a ruby laser. This technology was developed by Epperson et al. of Western Electric Co. in 1966 [9]. Many kinds of processing technology have subsequently been put into practice, and in the last five years, laser processing using high-power CO_2 lasers has spread rapidly into manufacturing industries.

In this chapter, interactions between laser beams and materials, equipment for processing, types of laser processing technology, and future trends in laser processing will be described.

2.2 Fundamentals of laser-beam processing

Laser-beam processing is mainly performed by thermal action. This is based on the rapid temperature rise in the limited area of the material surface irradiated by the focused laser beam. Laser beams can be applied in various kinds of material processing by controlling the power, power density and irradiation time, etc.

The laser beam is a form of coherent light, that is, of narrow spectral band width, of uniform phase, and with a parallel beam. Fine spots of extremely high power density can therefore be obtained on a workpiece by focusing with a lens. Using a focused beam of high power density, laser processing exhibits many features:

(i) The processing speed is very high and the heat-affected area very small compared with other machining methods, because the material of the workpiece is heated and evaporated instantaneously.
(ii) Workpieces need not be clamped tightly, because it is not necessary to apply machining force, so there is no risk of deforming or spoiling the workpiece. Consequently, it is easy to weld very thin foil and to machine very tiny parts.
(iii) Environmental considerations are of no consequence, since processing can be performed under ordinary atmospheric conditions, and also in various gaseous environments or in a vacuum.
(iv) No harmful X-rays are generated.
(v) A workpiece can be cut or welded behind any material which is transparent to energy of the wavelength of the laser beam.
(vi) Control of irradiation position and power are very simple.
(vii) It is easy to transmit the laser beam to several work stations located far from the laser oscillator.

2.2.1 Kinds of laser-beam processing

When a focused laser beam irradiates the surface of a workpiece, the temperature of the workpiece surface rises rapidly, as described above. As a result, localized heating, melting, and evaporation of workpiece materials occur, depending on the irradiation time. Figure 2.1 is a schematic diagram of the relation between temperature rise at the surface of an iron workpiece and irradiation time. On the right of the diagram, surface features at different temperatures are illustrated; the laser beam is projected from the left side. The surface is heated, melted, and finally evaporated with increasing irradiation time. With increasing power density of the laser beam, the temperature of the workpiece rises more rapidly. Drilling, cutting, and other material removal processes are performed mainly by evaporation. Welding, alloying, glazing, etc. are carried out in the temperature range above the melting point but below the vaporization temperature. Hardening occurs as a result of rapid cooling from the transformation temperature.

Laser processing has been classified according to the fundamental processing mechanisms, as shown in Table 2.1. Cutting, drilling, micro-machining, and cleaning are classed as material removal processes. Micro-machining includes trimming, marking, mask repairing, scribing, dynamic balancing, and tunning. Cleaning is the removal of contaminants such as oil films or gas adhering to the surface, by heating. Material consolidation processes include welding, padding, soldering, brazing, and cladding. Surface treatment includes hardening, annealing (metals, semiconductors), alloying, glazing, and coating, including laser CVD or PVD. New applications will soon be

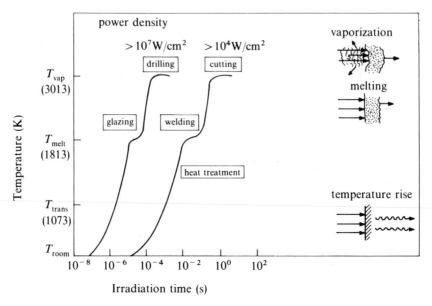

FIG. 2.1. Schematic diagram of temperature rise of iron surface irradiated by focused laser beam

developed throughout industry. Crystal growth, wiredrawing, cleaving, laser-assisted machining, laser-enhanced etching, and laser-induced chemical reaction are among the latest developments.

2.2.2 Interactions of laser beams with materials

(1) Absorption of light

Laser light irradiating the surface of a workpiece is partly absorbed; the rest is reflected or scattered. According to Hagen and Rubens's relation [10], the reflectance, R (%), of a metal surface for light of wavelength longer than 10 μm is expressed approximately by

$$R \approx 1 - 2\sqrt{(v/\sigma_0)} \qquad (2.1)$$

where σ_0 is the electrical conductivity of the metal ($\Omega^{-1}\text{m}^{-1}$) and v is the frequency of the light wave (Hz). Absorption of visible light, (that is, light of wavelength less than 10 μm) at the metal surface mainly causes interactions between the electric field components of the light and the electric dipole moments of the metal atoms, or absorption by conductive electrons accelerated in the electric field of the light. These phenomena are very complex. On the assumption that light scattering is very small and transmission through

TABLE 2.1
Classification of laser processing

Process	Method	Applications
Removal	Cutting	Metal, cloth, composite material, plastics, rubber, wood, glass, paper
	Scribing	Ceramics, semiconductor
	Drilling	Rubber, metal, diamond, rock, sapphire, plastics, ceramics
	Micro-machining	Triming, mask-repairing, balancing, tuning, marking
	Cleaning	Removing defects or strains
Consolidation	Welding	Metals, ceramics, micro-welding
	Sputter deposition	For hot evaporation source
	Brazing	
	Soldering	
Surface treatment	Hardening	Surface, partial quenching
	Tempering	Partial annealing
	Glazing	Changing to amorphous structure
	Alloying	Partial alloying
	Annealing	Recrystallization of semiconductor crystal after ion implantation
New process	Crystal growth	Sapphire crystal, Si-ribbon crystal
	Wiredrawing	Optical fibre
	Breaking	Cleaving glass or rock by thermal stress
	Drying	Drying ink
	Laser-assisted	Combined processing with laser and other method

metal is negligible, the following equation is obtained:

$$R + A = 100\% \tag{2.2}$$

where A is the absorbance (%). In Fig. 2.2 [11], the dependence of reflectance on wavelength is shown for several metals. It can be seen that reflection of infra-red light at metal surfaces is very high. Reflectances of metals for Nd–YAG laser light (1.06 μm) and CO_2 laser light (10.6 μm) are given in Table 2.2 [12], but these are values for ideal metal surfaces. In practice, oxide films

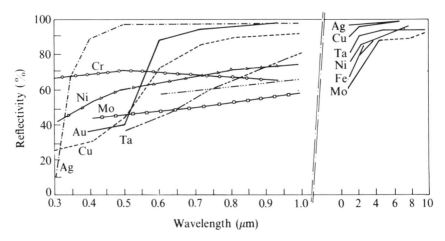

FIG. 2.2. Wavelength dependence of reflection at metal surfaces [11]

or adsorbed gas layers are encountered, and residual stress will exist at the metal surface. Reflectances at real metal surfaces will be different from the values calculated for ideal surfaces. However, there are very few data for reflectance at real metal surfaces. Such data would be of considerable practical value.

It is known that absorption characteristics correspond closely to the temperature gradient on metal surfaces in the initial period of CO_2 laser-beam irradiation, and that high absorption at the beginning of irradiation effectively improves processing. The variation in temperature rise rate, TRR (K/s),

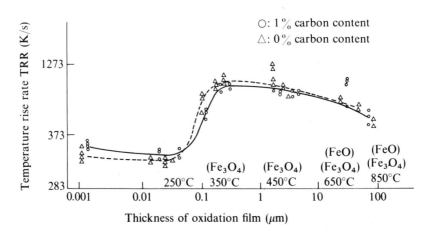

FIG. 2.3. Variation of absorption, as measured by TRR, with thickness of oxidation film on an iron surface [13]

TABLE 2.2
Reflectance of metals [12]

Metal	Reflectance (%)	
	Wavelength 0.9–1.1 μm	Wavelength 9–11 μm
Au	94.7	97.7
Pt	72.9	95.6
Ag	96.4	99.0
Al	73.3	96.9
Cu	90.1	98.4
Fe	65.0	93.8
Ni	72.0	95.6
Zn	49.0	98.1
Mg	74.0	93
Cr	57	93
Mo	58.2	94.5
W	62.3	95.5
V	64.5	92
Te	49.5	78
Ta	78.5	94
Sn	54.0	87.0
Si	28	28
Steel (1%C)	63.1	92.8–96
Constantan	72.4	94.2
Graphite	26.8	59.0

during the initial stage of laser-beam irradiation with the thickness of oxide film on an iron surface is shown in Fig. 2.3 [13]. At an oxide film thickness of 0.1 μm, a sudden increase in laser-beam absorption can be observed. However, if the oxide film becomes thicker than 2 μm, TRR decreases gradually, because of the low thermal conductivity of the oxide layer. The existence of a thin metal oxide film assists absorption of a CO_2 laser beam. There is no marked dependence of temperature rise on the carbon content of iron substrates.

Absorption also varies with surface roughness [13], as seen in Fig. 2.4. A rapid temperature rise is observed at a surface roughness around 0.03 μm. Here, absorption increases with increasing carbon content of the iron. TRR

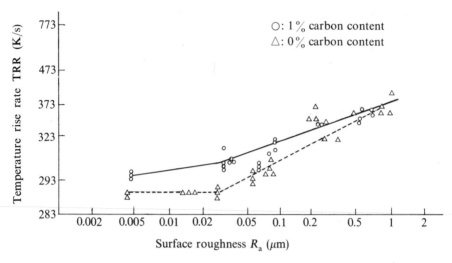

FIG. 2.4. Effect of surface roughness of iron on absorption as measured by TRR [13]

(K/s) is given by the following expression:

$$\text{TRR} \propto R_a^n \qquad (2.3)$$

where, R_a (μm) denotes the average surface roughenss and n is an experimentally derived constant. The value of n for 0 per cent carbon content is about 0.5, and for 1 per cent, about 0.3. This correlates with measured values of absorbance A.

The laser beam is absorbed by defects left on the surface by previous machining processes. Figure 2.5 shows the relation between TRR and surface strain as measured by half-value widths of X-ray curves diffracted by the (100) lattice planes of the Fe crystals [13]. It is seen that absorption increases with residual strain. The experimental results conform well with the theoretical relations between the stress relaxation time of materials and strain. The points for annealed specimens are clustered within a small range, but around a line showing the trend of this relation.

(2) Temperature rise

As described above, the surface temperature of a workpiece irradiated by a focused laser beam rises quickly. Table 2.3 shows the surface temperatures of highly heat-resistant (heat insulating) ceramics irradiated by a focused CO_2 laser beam of 80 W [14]. These were measured with a special micropyrometer. Some ceramic materials reach temperatures above the boiling point. These results indicate that nearly all materials can be melted and

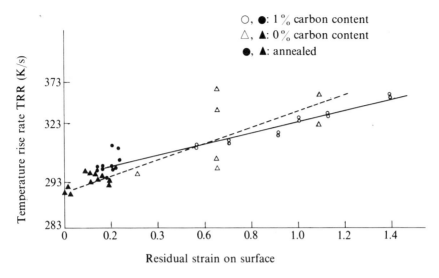

FIG. 2.5. Dependence of absorption, as measured by TRR, on surface strain [13]

TABLE 2.3
Evaporation temperatures of laser-irradiated refractory ceramics [14]

Materials	Temperature (K)	
	Measured	Vaporization
Zirconia (black)	4973	4573
Hafnia	4373	—
Zirconia (7%)	4373	4573?
Alumina	3773	3253 ± 69
Sapphire	3953	3253

vaporized by laser-beam irradiation. In other words, a laser beam has the ability to process thermally all kinds of material.

The distribution of temperature rise at the laser-beam irradiation point may be simulated by a model of heat flow to a semi-infinite body, as shown in Fig. 2.6. The temperature rise distribution is calculated for a heat source moving at velocity v (m/s) with Pittaway's equation [15], which is derived from Jaeger's formula [16]. The distribution of surface temperature rise is

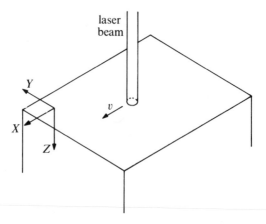

FIG. 2.6. Model for simulation of temperature rise

given by

$$\frac{T\lambda r_0}{Q_0} = \frac{1}{\pi^{\frac{3}{2}}} \int_0^{\frac{\pi}{2}} \exp\left\{\left[\left(m-\frac{1}{\alpha}\right)^2 + \beta^2\right]\sin^2\mu - \frac{\tan^2\mu}{\alpha^2} - (m^2+\beta^2)\right\} d\mu \quad (2.4)$$

In this equation, the power density distribution of the beam is assumed to be Gaussian; r_0 (m) denotes the radius of a focused beam spot as measured by the standard deviation of the Gaussian intensity distribution; λ (W/m K) is the thermal conductivity; Q_0 (W) is the laser power; β and m are given by $\beta = x/r_0$, $m = y/r_0$, where x is the distance from the beam centre in the X direction and y is the distance in the Y direction. The temperature, T (K), at the centre of the beam on the workpiece is given by

$$\frac{T\lambda r_0}{Q} = \frac{1}{\pi^{\frac{3}{2}}} \int_0^{\frac{\pi}{2}} \exp\left(\frac{\sin^2\mu - \tan^2\mu}{\alpha^2}\right) d\mu = 4\kappa/vd \quad (2.5)$$

where κ is the thermal diffusivity, $\lambda/\rho c$ (m²/s), and μ is defined by $m_t - m = \tan^2(\mu/\alpha)$, where m_t is the position of the beam at time t.

Substituting $v = 1.88$ cm/s, $Q_0 = 10$ W, and $r_0 = 0.1$ mm in eqn. (2.4), the temperature distribution for sapphire crystal may be calculated, and the results obtained are shown in Fig. 2.7. The stress is calculated on the basis of the thermal expansion and the temperature rise. The abscissa m denotes position in the direction of travel of the beam and the parameter β denotes the position in the direction along the surface perpendicular to the travel. In this case, room-temperature values of specific heat, absorption coefficient, thermal conductivity, thermal diffusivity, and thermal expansion coefficient have been used. In practice these values change with temperature, so if the correct temperature distribution on the workpiece is required, they must be corrected

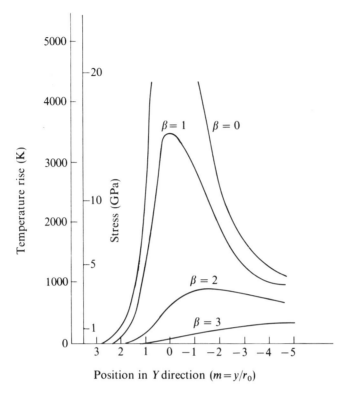

FIG. 2.7. Temperature and stress distributions on sapphire irradiated by a laser beam [15]

to the exact values at high temperature. Precise data for absorption coefficient, thermal conductivity, and thermal diffusivity at temperatures near the melting point are not fully available.

A new apparatus to measure absorption coefficient, specific heat, thermal conductivity, and thermal diffusivity has been developed for a CO_2 laser beam. It will be necessary in the near future to consolidate a data base on these thermal and optical properties, especially at temperatures near to melting point. Figure 2.7 presents only a rough estimate of the results, but from this it is seen that the surface temperature of the workpiece becomes extremely high within a short irradiation time, and that the area affected by heat is very narrow.

(3) Melting and vaporization

Cohen derived an equation to estimate the depth of melting by laser irradiation [17]. Assuming that heat flows one-dimensionally from the heated surface to the melt in a semi-infinite body, and that the thermal constants do not change during heating and melting, the depth of melting can be expressed approximately by

$$S_{max} \approx \frac{0.16(H^2 t_e - H^2 t_m)}{\rho L_m H} \quad (m) \qquad (2.6)$$

where H (W/m^2) is the intensity of stationary input heat flow, t_e (s) is the time from the start of laser irradiation to vaporization, t_m (s) is the time from the start of laser irradiation to melting, ρ is the density of the material (kg/m^3), and L_m (J/kg) is the latent heat of fusion. $H^2 t_e$ and $H^2 t_m$ are constants depending on the type of work material. These values have been calculated by Cohen.

The depth S (m) of a hole produced by vaporization is expressed approximately by the following relation derived from Landau's equation [18]:

$$S \approx \frac{H(t - t_m)}{\rho(L_v + CT_v)} \qquad (2.7)$$

where t (s) is the irradiation time, L_v (J/kg) is the latent heat of vaporization, T_v (K) is the vaporization temperature, and C (J/kg K) is the specific heat capacity of the material. If the melt on the surface is blown off immediately, the boring speed, or recession speed, of the hole bottom V (m/s) is expressed approximately by

$$V \approx \frac{H}{\rho\{L_m + C(T_m - T_0)\}} \qquad (2.8)$$

where T_0 is room temperature. If phase change is considered, the result obtained should be modified. Miyazaki and Taniguchi [19] have proposed the use of corrected melting and boiling points.

(4) Residual strain and structure change

Cross-sections of the molten area of a workpiece irradiated by a moving laser beam are illustrated in Fig. 2.8. When a laser beam irradiates the workpiece surface, melting of the surface occurs instantaneously, the melted area spreads quickly, and the melt begins to vaporize. When vaporization occurs, the melt is displaced by the vapour pressure and a small initial hole is produced. The laser beam then penetrates more deeply through the hole. The transverse section (a) shows that the melt is pushed to both sides, and the longitudinal section (b) that it is pushed backwards. If irradiation is stopped, or if the laser beam moves forward, the melt begins to solidify or recrystallize.

The residual strain around the hole made in a silicon single crystal by ruby laser-beam irradiation has been observed by X-ray topography [20]. By this method, it is possible to detect a strain of $1 \times 10^{-5} \sim 5 \times 10^{-6}$. The sensitivity

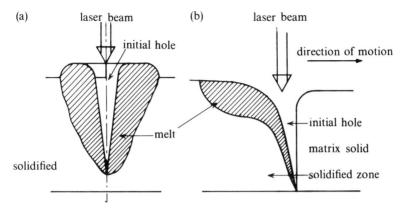

FIG. 2.8. Behaviour of work material irradiated by a laser beam. (a) Transverse section. (b) Longitudinal section

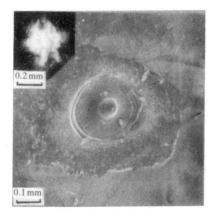

FIG. 2.9. SEM photograph of a hole drilled in Si single crystal by a 0.6 J ruby laser pulse, and X-ray topograph of distorted area (upper left) [20]

of the method varies, depending on the lattice plane participating in X-ray diffraction. Figure 2.9 shows a SEM (scanning electron microscope) photograph of the hole made in silicon crystal by one ruby laser pulse of 0.6 J. Solidification of the melt is observed at the hole surface and periphery. The X-ray topograph of the hole shows the distorted area as white, because in this area the X-rays are not diffracted normally and do not sensitize the film. The experimentally obtained relation between the diameter of the hole, measured at the surface, and that of the distorted area is shown in Fig. 2.10. The diameter of the distorted area is approximately 1.5 times that of the hole. The

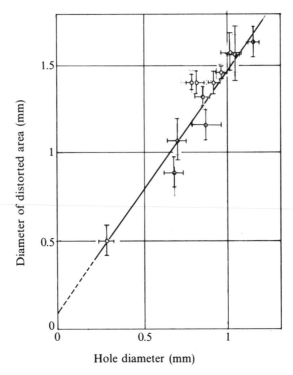

FIG. 2.10. Relation between diameter of hole drilled in Si single crystal by ruby laser pulse and diameter of distorted area [20]

worked area on the Si single crystal becomes a polycrystalline structure. The state of residual distortion on the Si crystal caused by laser-beam irradiation is not simple. With metals, which are generally polycrystalline, the state of distortion caused by laser-beam irradiation is more complex than that of an Si crystal, but the size of the distorted area seems to be nearly the same.

Various kinds of change in structure or composition occur during laser-beam irradiation of materials [21]. A cross-sectional view and a diagram of the irradiated area of iron (1 per cent C) are shown in Fig. 2.11. In this case a CO_2 laser beam of about 150 W was focused with a ZnSe lens of about 190 mm focal length. The structure was analysed with a micro X-ray diffractometer of 0.1 mm spot diameter, using Cu Kα X-rays. The cross-section shown in Fig. 2.11 was polished and etched with 5 per cent nital etchant. Hardness was measured with a Vickers microhardness tester. At the surface of the iron plate, a shallow crater is formed. The size of the crater increases in proportion to irradiation time. The surface of the crater bottom is covered with an oxide film about 0.1 mm thick. There is a white layer (I) about

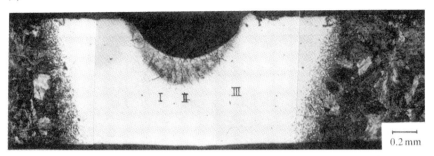

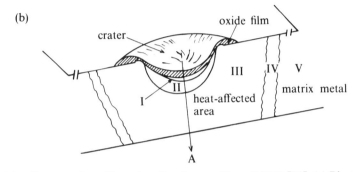

FIG. 2.11. Cross-section of laser-irradiated area of iron (1%C) [21]. (**a**) Photograph at centre of crater. (**b**) Diagrammatic representation

30 μm thick whose surface is covered with an oxide film, and below it a dark, crescent-shaped area (II). A white structure (III) which is difficult to etch surrounds them. The diameter of the white area is about 2.4 mm on the upper surface and about 2.0 mm on the reverse side. The structure and hardness distributions along the arrow A are shown in Fig. 2.12. The Vickers microhardness of structure (I) varies from 700 to 800. It has a mixed structure of martensite (α') and austenite (γ), with some parts of α-iron. In area (II), the hardness decreases to about 350 to 550, and the structure consists only of α-iron. The structure of area (III) is mainly martensite and its hardness reaches about 850 to 950°. This area is apparently hardened. When the area is etched for more than 20 s, a needle-shaped structure appears.

2.3 Equipment for laser-beam processing

2.3.1 Laser oscillator or source

Many kinds of laser oscillator have been developed. Lasers may be classified into three groups: solid, gas, and liquid, according to the phase of the laser

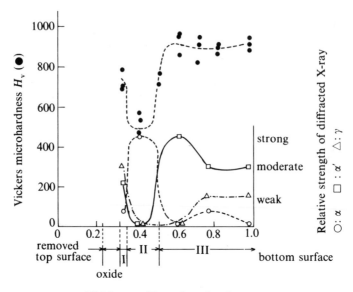

FIG. 2.12. Structure and hardness distributions of laser-irradiated area along arrow A in Fig. 2.11 [21]

medium. The ions, atoms, and molecules of this oscillating medium take part directly in laser action (Table 2.4). The fundamental structure of laser equipment is shown in Fig. 2.13; it consists of a laser medium, an excitation energy source, and a Fabry–Perot resonator. When the oscillating medium is excited, the number of atoms, ions, or molecules in the upper excited state increases rapidly. As a result, the state of inverse population is realized for an instant. An electric discharge, electron beam, visible light, ultra-violet light, chemical reaction, heat, etc., may be employed as sources of excitation energy.

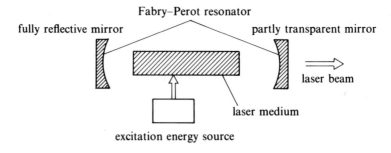

FIG. 2.13. Fundamental structure of laser equipment [48]

TABLE 2.4
Main types of laser used for thermal processing

Medium	Laser	Wavelength (μm)	Output	Power
Solid	Ruby (Cr^{3+})	0.694	Pulse	50 J per pulse
	YAG (Nd^{3+})	1.064	CW pulse	600 W 100 J per pulse
	Glass (Nd^{3+})	1.064	Pulse	200 J per pulse
	Alexandrite	0.7–0.818	Pulse	Tens of watts (mean value)
Gas	CO_2	10.6	CW pulse	20 kW
	Ar^+	0.488, 0.514	CW	100 W
	Excimer (ArF, KrF)	0.19–0.351	Pulse	10 W (mean value)
	He–Cd	0.33, 0.44	CW	10 W
Liquid	Dye (Rhodamine 6)	0.34–1.175	Pulse	Hundreds of watts (mean value)

When an excited particle spontaneously emits laser spectral light and returns to ground level, the emitted light induces simultaneous light radiation by other excited particles. This phenomenon is called induced emission. Emitted light is stored within the resonator for an extremely short time and collimated. Part of the stored light emerges through the partly transparent mirror of the resonator, as the laser beam. The excitation and induced-emission processes are repeated continuously.

The main types of laser equipment used for thermal applications are as follows. Ruby, Nd–YAG, Nd–glass, alexandrite, etc., belong to the solid-state laser group. He–Ne, CO_2, Ar ion, excimer, He–Cd, etc., are classified as gas lasers. A representative liquid laser is the dye laser. New high-power lasers such as the chemical laser, metal vapour laser, CO laser, etc., have been studied. These lasers will be introduced for new practical applications in the future.

The main types of equipment used in practice for materials processing are the Nd–YAG and CO_2 lasers. Some glass and ruby lasers are also used. The ruby laser was used at first, but today the Nd–YAG laser is used more widely in industry because of the good thermal characteristics of YAG crystal. The alexandrite laser will be used in materials processing in the near future.

(1) YAG laser

The laser medium referred to as YAG consists of a $Y_3Al_5O_{12}$ isometric crystal containing about 1 per cent of Nd^{3+} ions. This crystal was developed by Geusic *et al.* in 1962. Its thermal conductivity is ten times that of glass. Continuous oscillation is possible with YAG. The Nd^{3+} ions form the oscillating medium, which shows the four-level laser action typical in solid-state lasers. Four energy levels designated E_0 to E_3 participate in the oscillation [22]. The energy levels and a laser transition of Nd^{3+} ions are shown in Fig. 2.14. Even if the concentration of Nd^{3+} ions in the crystal increases, the spectrum of the oscillating light does not become broad, because the valency and ion radius of Nd^{3+} are not very different from those of Y^{3+}. The crystal oscillating medium component of this laser is in the form of a long, slender rod, both end surfaces of which are precisely polished in parallel; the side is also finished to a mirror surface.

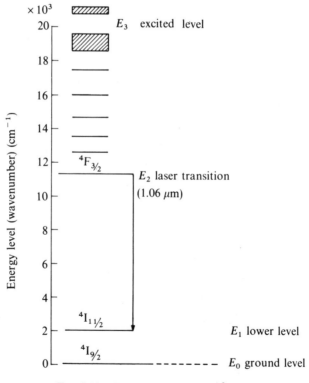

FIG. 2.14. Energy levels of Nd^{+3} [22]

FIG. 2.15. Photograph of interior of a YAG laser oscillator [32]. The major axis of the reflecting cavity is 30.6 cm long

Figure 2.15 is an interior view of an Nd–YAG laser oscillator. Nd^{3+} ions are excited with the light from an electric arc lamp or a spiral flash lamp. The former is used for oscillating a continuous wave (CW) and the latter for pulsed light. The YAG rod and excitation lamp are installed in the cavity of a reflecting mirror [23]. The shape of the cavity is an elliptic cylinder or an ellipsoid of revolution, which is sometimes formed as double ellipsoid. A cylindrical reflecting cavity is usually employed for the flash lamp. The reflecting mirror cavity concentrates light from the lamp onto the rod (Fig. 2.16). A representative construction of a Nd–YAG laser is shown in Fig. 2.17. In addition to the reflecting cavity housing the lamp and rod, a beam shutter, a Q-switch, and a mode selector are set inside the resonator to control the oscillating mode. Oscillating modes and their main applications are listed in Table 2.5.

(2) CO_2 laser

The CO_2 laser is used in materials processing as a high-power laser. The oscillator is of compact size and high energy efficiency. The laser medium is a mixture of three or four gases: CO_2, N_2, He, and CO. The oscillating medium is the CO_2 molecules, but N_2 molecules assist their excitation. CO_2 molecules

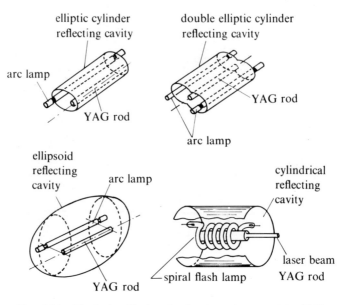

FIG. 2.16. Typical gold-plated reflecting mirror cavities [49]

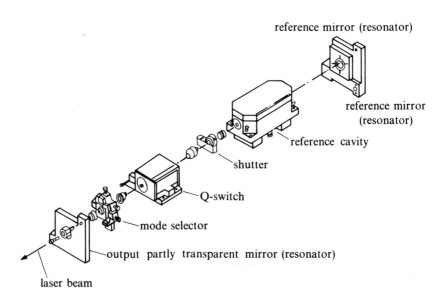

FIG. 2.17. Practical construction of YAG laser oscillator [49]

TABLE 2.5
Oscillating beam mode of YAG laser and applications [32]

Excitation	Output			
	Continuous	Pulse		
	CW	Quick repetition pulse	Slow repetition pulse	Giant pulse
Lamp	Arc lamp	Arc lamp	Flash lamp	Flash lamp
Q-switch	Not used	Acoustic optic rotating mirror	Not used	Electro-optic
Pulse width		150–200 ns	0.1–20 ms	0.1–10 ms	10–40 ns
Repetition frequency		50 kHz	100–200 pulse/s	1–50 pulse/s	1–50 pulse/s
Peak power		10–20 kW	1 kW	10–20 kW	A few MW–a few W
Mean power	300–400 W	100 W	300–400 W	300–400 W	
Applications	Welding, soldering	Trimming, cutting, scribing, marking, annealing circuit correcting	Drilling, cutting, trimming, balancing, marking	Welding, cutting, soldering, brazing, heat treatment	Measurement, scientific application

at lower levels quickly return to ground level by collision with He atoms [24]. The presence of CO in the laser medium prevents an increase in dissociation of the CO_2 during discharge. Generally, the oscillating medium in industrial CO_2 lasers is excited by means of a d.c. glow or high-frequency discharge. There is also a laser which employs electron-beam sustained excitation.

Figure 2.18 is a model of the vibration energy levels involved in laser oscillation of a CO_2 molecule. This shows that N_2 molecules are first excited to the upper level by the d.c. glow discharge. The excited energy of the N_2 molecules is then transferred to CO_2 molecules by elastic collision, and the N_2 molecules return from the upper energy level to the ground state. Then excited CO_2 molecules simultaneously radiate infra-red light of wavelength 10.6 and 9.4 μm by induced emission and also return to ground level.

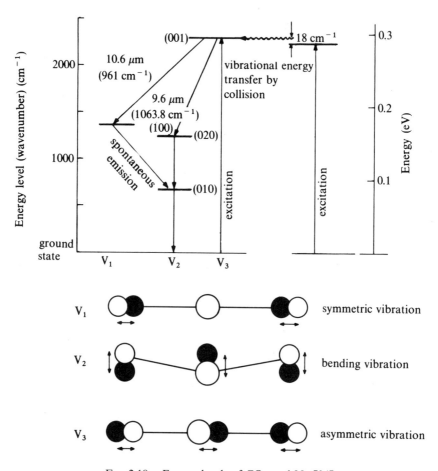

FIG. 2.18. Energy levels of CO_2 and N_2 [24]

The CO_2 molecule consists of a carbon atom bonded to two oxygen atoms, one on either side. The upper energy level which is shown by (001) in Fig. 2.18 corresponds to asymmetric vibrational motion (V3) of the CO_2 molecule. The lower two levels correspond to symmetric and bending vibrations, respectively.

The construction of CO_2 laser oscillators for materials processing may be classed into coaxial and cross-flow types according to the directional relation of the axes of the optical resonator and gas flow [25]. The cross-flow type is subdivided into orthogonal biaxial and orthogonal triaxial types according to the directional relation between gas flow and discharge: parallel in the former and orthogonal in the latter. The structure of the three types is illustrated in Fig. 2.19.

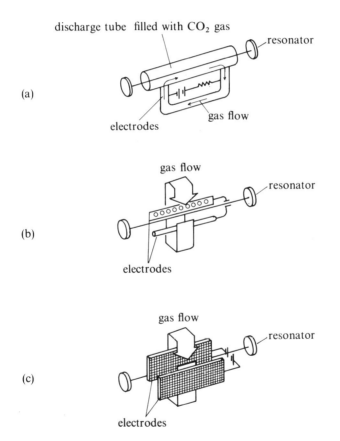

FIG. 2.19. Diagrams of three types of CO_2 gas laser oscillator [24] (a) Coaxial type. (b) Orthogonal biaxial type. (c) Orthogonal triaxial type

The coaxial type is the most popular laser oscillator, with glass discharge tubes. A beam of axially symmetric energy distribution, especially of the Gaussian mode (transverse mode TEMoo), can be obtained easily. The output power obtained per unit discharge length (m) is only about 50 W. Generally, in order to increase the output power it is necessary to increase the number of excited gas molecules in the discharge tube between the resonators. This is generally achieved by increasing the gas pressure, or by increasing the gas flow velocity between the resonators. Rapid gas flow, over 100 m/s, and elongation of discharge length or resonator distance are applied in practice. In the high-power laser oscillator, a long-distance resonator of multipath structure, furnished with a deflected optical path is employed, as shown in Fig. 2.20. However, in this structure it is difficult to align the optical path of the resonator and the deflecting mirrors. Furthermore, the distance between the discharge electrodes is rather long, so a high voltage is required for stable glow discharge, and then it is necessary to take steps to avoid breakdown.

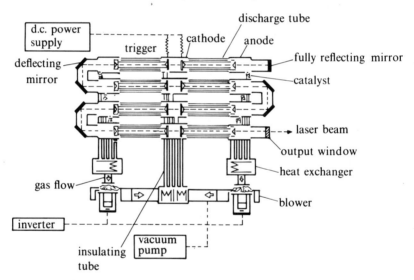

Fig. 2.20. Configuration of a 5 kW fast-axial-flow CO_2 laser oscillator [26]

As regards the orthogonal biaxial type, a large cross-section of discharge region is easily achieved, so the length of the laser oscillator in the resonator direction can be less than that of the coaxial type. However, fast gas flows of over 100 m/s and a homogeneous and stable discharge over a large cross-section are required, so the gas circulation technology becomes very important. To obtain laser energy effectively from the large discharge area, or from the large volume of excited gas, a multipath resonator or an amplifier system is generally used.

Of the three types, the orthogonal triaxial laser has the shortest distance between the electrodes, and requires the least gas flow velocity, 30–80 m/s. This type has a low discharge voltage and a low gas flow, but the operating pressure is rather high. In this oscillator, therefore, a stable and homogeneous d.c. glow discharge is the key. Glow discharge excitation sustained by an electron beam and a high-frequency discharge, or high-frequency silent discharge excitation are also employed.

The first industrial laser oscillator on the market was a coaxial type with a low gas flow rate. This reasonably sized laser oscillator was produced with a stable output and high reliability at an output power less than 250 W. However at more than 500 W output the oscillator was difficult to apply in practice, because of lack of stability and reliability. To solve these problems, the fast gas flow coaxial type was studied by the Welding Institute in England and the technology developed was taken up by British Oxygen Co., to market the first kilowatt class laser. This laser oscillator is now made by Control Laser Co. (USA). Recently a fast gas flow coaxial laser with output power 0.5–1 kW with TEMoo single mode has been marketed by Rofin Sinnal Co. (West Germany), incorporating a Roots-type blower operated at high gas pressure. A diagram of a highly efficient 5 kW fast axial flow CO_2 laser oscillator developed by Hitachi Co. (Japan) is given in Fig. 2.20 [26]. The total energy conversion efficiency of this laser oscillator reaches 26 per cent. Hitachi are now concentrating their efforts on developing a 20 kW laser oscillator.

An experimental orthogonal biaxial laser was developed by United Technologies Research Center (USA) and a maximum output power of 27 kW was achieved. The first product of this type was manufactured by Toshiba Co. (Japan), with an output power of 5 kW. The gas flow channel and the detailed construction of the electrodes etc., are illustrated in Figs. 2.21 and 2.22 respectively. The characteristics of the oscillator using stable and unstable resonators are shown in Figs. 2.23 [27] and 2.32. The energy conversion efficiency at maximum output power was about 15 per cent in the case of the stable resonator, but in the unstable resonator it was only 11 per cent.

The orthogonal triaxial laser oscillator was first developed at the Culham Laboratory (England). Spectra Physics (USA), AVCO (USA), and Mitsubishi Electric Co. (Japan) put this type of oscillator on the market successively. Culham and SP employed a d.c. glow discharge excitation. AVCO employed a d.c. discharge, controlled by electron-beam excitation, but the system was considered to be unsuitable for industrial applications because of its complex structure and difficult maintenance.

Mitsubishi has marketed CO_2 laser oscillators of about 10 kW and is currently developing a 20 kW laser oscillator [28]. This oscillator incorporates a new excitation technique, called silent-discharge-assisted d.c. glow excitation (SAGE). This kind of oscillator has high beam quality, high efficiency, high reliability, and great durability, and the overall set-up is

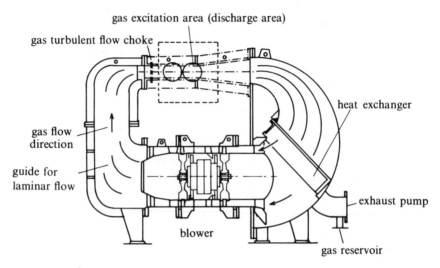

FIG. 2.21. Diagram of an orthogonal biaxial laser and gas flow channel [27] (Toshiba Co.)

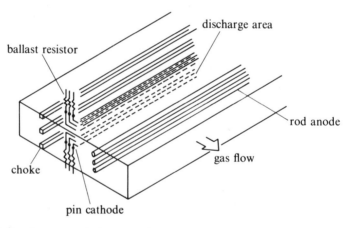

FIG. 2.22. Structure of electrode of orthogonal biaxial CO_2 laser oscillator [27] (Toshiba Co.)

compact. The cross-section of the gas flow channel perpendicular to the optical axis and details of the electrodes for SAGE are illustrated in Figs. 2.24 and 2.25, respectively. The electrodes consist of a dielectric electrode, pin cathodes, and anode plates. Silent discharge is especially effective in increasing the discharge input power per unit discharge volume, exciting a large volume uniformly at a gas pressure of about 100 torr (1.33 kPa), and

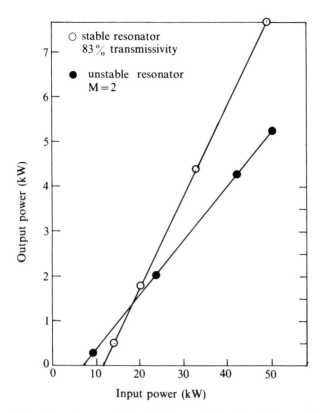

FIG. 2.23. Oscillation characteristics of orthogonal biaxial CO_2 laser oscillator. M is the ratio of the outside to the inside diameter of annual mode (see Fig. 2.32). (Toshiba Co.)

oscillating a pulsed output. The gas pressure dependence of the 20 kW laser output is shown in Fig. 2.26, which indicates that the efficiency will reach about 17 per cent. Representative CO_2 lasers on the market are listed in Table 2.6.

2.3.2 Focusing and transmission of the laser beam

(1) Focusing of the beam [29]

To achieve successful laser-beam processing, it is necessary to direct the focused laser beam onto the appropriate point on the workpiece. Focusing is effected using a lens or a mirror. Materials for the focusing optics have to be selected in accordance with the wavelength of the laser.

If a Gaussian-distribution beam of radius r_0 (m) corresponding to $1/e$ is

TABLE 2.6
Main types of CO_2 laser on the market [48]

Structure		Output power	Maker	Excitation	Resonator[†]
Coaxial flow	Slow gas flow	Up to 1 kW	Photon Sources Co. Coherent Co. NEC Shimada Sci. Inst.	D.c.	ST
	High-speed flow	2–5 kW	Control Laser Co. Hitachi Co.	D.c.	ST
		2.5–5 kW	Osaka Trans. (5 kW)		Amp (Osaka)
		0.5–1 kW	Rofin Sinar Co. National Research Inst. (Japan) Osaka Trans. Co.	D.c.	ST
Cross-flow	Biaxial	27 kW	UTRC	D.c, r.f. (assisted)	Amp
		1–5 kW	Toshiba Co.	D.c.	ST US (5 kW)
	Triaxial	1–5 kW	Culham Laboratory Spectra Physics Co. Hitachi Co.	D.c.	ST
		1 kW			
		15–20 kW	AVCO	D.c., electron beam (controlled)	US
		1–5 kW		Silent discharge	ST
		5–10 kW	Mitsubishi Electric	SAGE	US

[†] ST = stable resonator, US = unstable resonator, Amp = combination of oscillator and amplifier

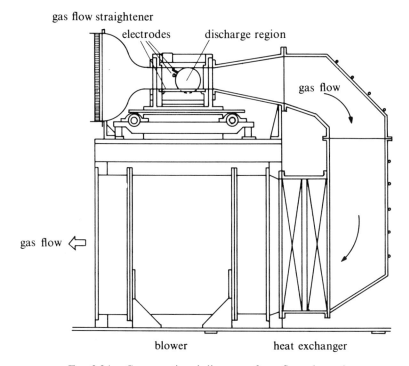

FIG. 2.24. Cross-sectional diagram of gas flow channel (Mitsubishi Electric Co.) [28]

focused, using a lens of focal length f(m) and a diameter sufficiently larger than that of the beam, then the radius r_m (m) of the focused beam spot and the distance d_m (m) from the lens to the position of minimum spot size (sometimes called the beam waist) are given by

$$d_m = f + \frac{(d_1-f)^2}{[(d_1-f)^2+f_{\tilde{H}}^2]^{\frac{1}{2}}} \tag{2.9}$$

$$r_m = \frac{fr_0}{[(d_1-f)^2+f_{\tilde{H}}^2]^{\frac{1}{2}}} \tag{2.10}$$

where $f_{\tilde{H}} = nf\lambda_w/r_0$ and n is the refractive index. The spot radius at the focal length of the lens is given by

$$r = \frac{f\lambda_w}{\pi r_0} \tag{2.11}$$

where λ_w is the wavelength of the laser light. It is found that the spot size

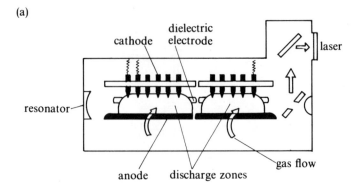

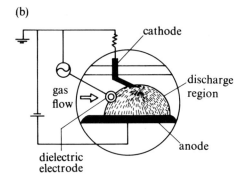

Fig. 2.25. Schematic diagram of SAGE electrode (**a**) and discharge (**b**) (Mitsubishi Electric Co.) [28]

decreases with increasing beam diameter and decreasing focal length or laser beam wavelength. For a multimode beam, the spot radius is given by

$$r = f\theta \qquad (2.12)$$

where θ is the divergence angle (rad).

The minimum aberration (A_{min}) of a lens of refractive index n is given by

$$A_{min} = \frac{n(4n-1)}{8(n-1)^2(n+2)f} \qquad (2.13)$$

That is, the shape of lens which gives the minimum aberration is determined by the refractive index of the lens material. The spot radius r, taking aberration into account, is given by

$$r = 0.61 f \lambda_w / r_0 + Y_{min} \qquad (2.14)$$

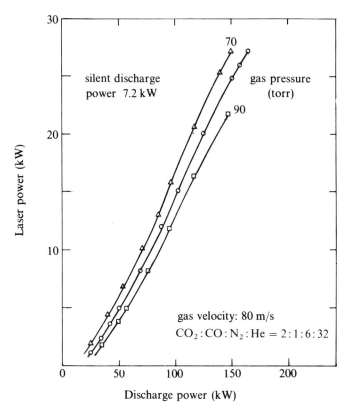

FIG. 2.26. Oscillating characteristics of SAGE orthogonal triaxial CO_2 laser oscillator (Mitsubishi Co.) [28]

and Y_{min} is given by

$$Y_{min} = A_{min}(r_0^3/2f) \qquad (2.15)$$

For a higher-power laser beam, reflection focusing systems are employed. Typical examples are shown in Fig. 2.27 [30]. A parabolic mirror is the best focusing system for a high-power laser beam.

(2) Transmission of the beam

It is necessary to transmit the laser beam from the oscillator to the machining table and to project it to the required position on the workpiece. Mirrors are used in the transmission system to bend and control the position of a CO_2 laser beam. Various beam transfer systems have been developed, as shown in Fig. 2.28. These are employed for cutting sheet materials. The systems are classified into beam-scanning, moving-table and a combination types. The

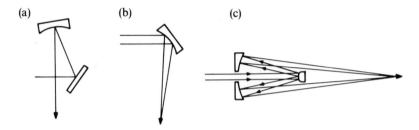

FIG. 2.27. Focusing systems using reflecting optics [30]. (a) Flat and concave mirrors. (b) Parabolic mirror. (c) Cassegrain mirror

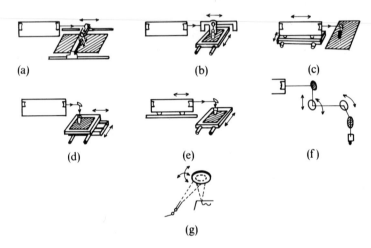

FIG. 2.28. Laser beam transmission and scanning systems [31]. (a) Moving lens. (b) Moving lens and workpiece. (c) Moving oscillator. (d) Moving workpiece. (e) Moving oscillator and workpiece. (f) Multi-joint robot arm. (g) Tilting mirror

type should be selected according to the size of the oscillator, size of work material, beam quality, beam scanning speed, and required machining accuracy. Recently, focusing and positioning systems with numerical control devices for simultaneous motion about five axes have been developed for cutting or welding three-dimensional workpieces.

A YAG laser beam can be transmitted by an optical fibre. General Electric Co. (USA) has succeeded in transmitting a peak power of 10 kW or 400 W CW power of a YAG laser beam of 1.8 cm diameter over a distance of 25 m with an optical fibre of 1000 μm core diameter [31]. NEC Co. (Japan) has successfully transmitted a 300 W YAG laser beam using a fibre of 400 μm core diameter [32]. Optical fibre may be very conveniently used to split a laser

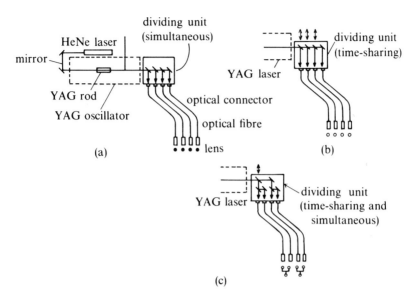

FIG. 2.29. Laser beam transmission and branching systems using optical fibre [48]

beam into several branches and to send them to several work stations, as seen in Fig. 2.29. Two beam-positioning systems are shown in Fig. 2.30.

2.3.3 Optical components of the high-power CO_2 laser

The optical properties of the optical components play a large role in processing with a high-power laser beam of wavelength 10.6 μm.

(1) Transparent components [33]

These comprise windows, beam splitters, lenses, etc. The materials used for the laser of 10.6 μm extreme infra-red light are different from those used for visible light. These materials are evaluated in terms of thermal fracture strength (f_T) and optical distortion (f_0). The former is an index of fracture strength with partial heat absorption and thermal expansion, while f_0 denotes the distortion of an image passing through the transparent components, caused by optical path or refractive index deviation due to inhomogeneous heating. These indices are defined as

$$f_T = \frac{\sigma_c \lambda}{\beta \alpha E} \qquad (2.16)$$

$$f_0 = \frac{\lambda}{\beta \chi} \qquad (2.17)$$

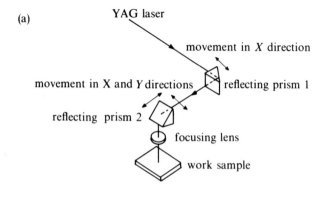
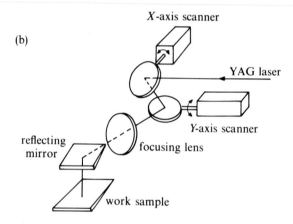

Fig. 2.30. Schematic construction of beam positioning systems [49]. (a) Table type. (b) Galvanometer type

values of σ_c, λ, α, β, χ are given in Table 2.7 for representative materials. Calculated values of these indices are listed in Table 2.8. In practice, ZnSe and KCl are mainly used. These are characterized by low absorption power, moisture resistance, transparency to the He–Ne laser beam, etc.

(2) Metal mirrors [34]

The main materials used in reflecting mirrors are copper, beryllium–copper and tungsten. These are materials of high thermal conductivity. Surfaces are finished by polishing or diamond turning, and are coated with a protective evaporated film of gold or silver. Reflectances of different materials are given in Table 2.9.

TABLE 2.7

Characteristics of transparent materials [33]

Material	Wavelength of transmitted light (μm)	Absorption coefficient β (m^{-1})	Refractive index[†] n	Thermal conductivity λ (W/m K)	Thermal expansion coefficient α (10^{-6}/K)	Young's modulus E (GPa)	Breaking stress σ_c (MPa)	Optical distortion parameter χ (10^{-6}/K)
Ge	1.8–23	1.2×10^{-4}	4.02	59	5.7	101.9	92.6	317
CdTe	0.9–30	2.5×10^{-6}	2.69	6	5.9	23.5	30.9	117
GaAs	0.9–18	8×10^{-5}	3.30	48	5.7	84.3	137.2	160
ZnSe	0.5–22	1×10^{-5}	2.40	18	8.5	66.6	54.9	67
NaCl	0.2–18	1.3×10^{-5}	1.52	6.5	44	40.2	3.9	−0.6
KCl	0.2–24	7×10^{-7}	1.47	6.5	36	29.4	4.3	−5.0
KBr	0.2–30	4.2×10^{-3}	1.54	4.8	42	26.5	3.3	−3.6

[†]For 10.6 μm

Table 2.8
Evaluation indices of transparent materials for 10.6 μm light [33]

Material	f_T	f_0
Ge	7.8×10^7	0.15×10^{-6}
CdTe	53.8×10^7	2.04×10^{-6}
GaAs	17.1×10^7	0.6×10^{-6}
ZnSe	17.3×10^7	2.68×10^{-6}
NaCl	0.11×10^7	83.3×10^{-6}
KCl	3.8×10^7	185.7×10^{-6}
KBr	0.34×10^7	0.03×10^{-6}

Table 2.9
Reflection characteristics of mirror at 10.6 μm [34]

Base material	Coating metal	Reflection coefficient (%)
Cu	–	99.2
Cu	Au	99.2–97.1
Cu	Ag	98.9–95.9
W	–	98
W	Au	99.1–97.8

2.3.4 Measurement of laser beam characteristics

Measurements of power and distribution of power density in the beam are described in this section. Precise measurement of the beam characteristics is necessary to obtain high quality and high reliability in processing.

(1) Measurement of laser power

The measuring methods employed in practice are thermal or optical–electrical methods. Many kinds of instrument for low power measurement are sold on the market. For the high-power beam, the problem is to materials with low absorption characteristics and which can withstand high power. If a high power density beam irradiates a detector, the surface will be damaged within a short time. The damage threshold of input power density on the surface of detectors used in practice is less than 500 W/cm². However, for practical use, a threshold of power density of 1–2 kW/cm² is necessary. A

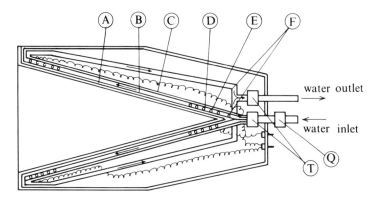

FIG. 2.31. Schematic diagram of internal structure of high-power laser calorimeter [35]. A = extinction surface, B = aluminium substantial cone, C = outer substantial cone, D, E = calibration heater, F = cooling water flow, T = quartz thermometer, Q = flow meter

typical structure designed to reduce the beam power density on the absorbing surface is shown in Fig. 2.31 [35]. This flow-type calorimeter employs a multiple reflection structure by which the laser energy is absorbed and removed by flowing coolant (water). As a result, the temperature of the coolant rises. Laser power is then measured by calculating the energy corresponding to the observed temperature rise of the coolant. Calibration is performed by means of the electric heater attached to the back of the absorbing surface.

(2) Measurement of beam mode [36]

The beam mode (profile of power density distribution) is determined by the structure of the resonator. Figure 2.32 illustrates the conceptual relation between type of resonator and beam mode. The convergence characteristics of the laser beam depend on the beam mode, beam diameter, and divergence angle, as already seen in Section 2.3.2 (1).

A beam mode measuring system has been developed as shown in Fig. 2.33. The detector is a HgCdTe device cooled to 77 K by liquid nitrogen in a Dewar. In this system the detector is basically scanned two-dimensionally along a cross-section of the beam. The HgCdTe detector, an attenuator, a rotary prism, and a beam splitter are mounted on a stage which is moved in the vertical direction by a screw driven by a stepping motor. The laser beam is scanned by a rotating prism in the horizontal direction; 200 scans are made and data for 200 points are sampled during each scan. A microprocessor as a host computer controls the motion of this apparatus, sets parameters, processes and operates signals, and displays the results obtained. The power

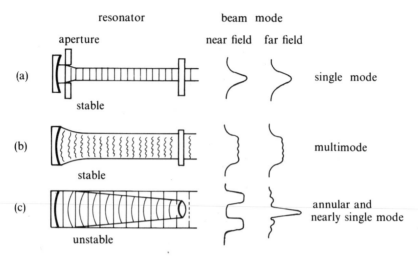

Fig. 2.32. Relation between beam mode and type of resonator [48]

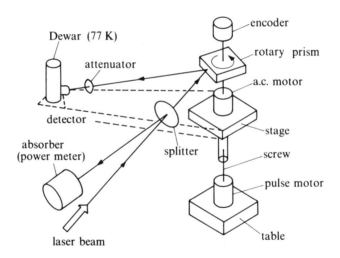

Fig. 2.33. Schematic diagram of beam profile measuring systems [36]

of the laser beam is reduced by a factor of 10^5 with a beam splitter. Values at 200×200 points over the cross-section of the 50 mm diameter beam are measured and stored in 8 bit IC memories. Using these measurements, the following data are calculated and displayed on a CRT with the microcomputer:

(i) the point of highest power density
(ii) the practical power density of the beam

(iii) the power density profile at any cross-section
(iv) contour lines of power density
(v) a three-dimensional plot of power density distribution
(vi) the mean power density.

Examples of a three-dimensional plot and power density profiles are shown in Figs. 2.34 and 2.35

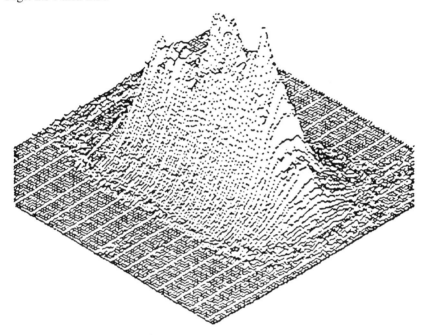

FIG. 2.34. Three-dimensional plot of power density distribution of the laser beam [36]. Total power = 2 kW, peak = 0.802 kW/cm^2

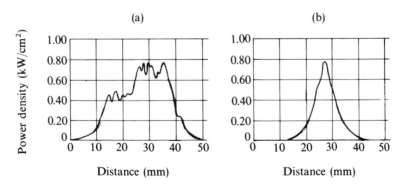

FIG. 2.35. Profiles of power density across a laser beam [36]

Generally, for convenience, power density distribution is observed by means of a crater made in polymethyl methacrylate (PMMA) by irradiation with a CO_2 laser beam. The depth of the crater increases with irradiation time and its outline shows a tendency to expand, but the boundary becomes indistinct. It is found that there is a power threshold for forming a crater on PMMA. The time dependence of the threshold is shown in Fig. 2.36. It is possible to estimate power density distribution from a crater made at the optimum duration of irradiation.

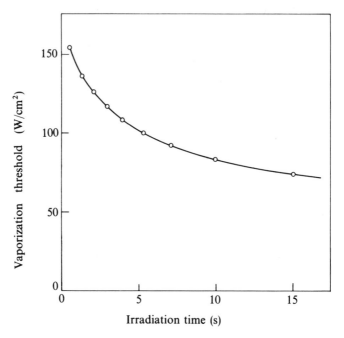

FIG. 2.36. Relation between irradiation time and power density threshold for vaporization of PMMA by CO_2 laser beam [36]

(3) Measurement of beam waist

The position of the minimum focused beam diameter is called the beam waist. To calculate the power density, measurement of the diameter of the beam waist is necessary. Generally, the waist diameter can be estimated by measuring the width of the trace made by fast scanning of a focused laser beam on a tilted PMMA plate. Steen has developed an apparatus for the measurement of spot sizes of focused high-power CO_2 laser beams [37]. The device, using a rotating wire, is capable of measuring the instantaneous size of a laser beam with an accuracy of about ± 0.02 mm.

2.4 Laser-beam machining

2.4.1 Drilling

Drilling may be performed using either a CW or a pulsed laser beam. The size of the thermally affected zone around a hole drilled by a pulsed laser beam is smaller than that drilled by a CW beam, and the thermal efficiency is also higher. The hole drilled by a single pulse is not very deep, so it is necessary to apply many repeated pulses. The depth and shape of the hole and the stock removal rate are strongly influenced by the focal length of the lens, the position of the focusing point, the energy per pulse, etc. It is important that the

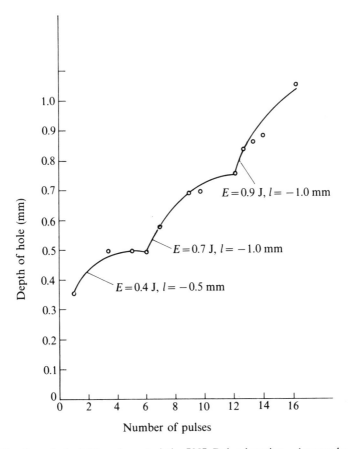

FIG. 2.37. Deep-hole drilling characteristics [38]. Pulse duration = 1 ms; pulse energy = E (J); distance of focal point from surface = l (mm)

optimum machining conditions should be assessed according to the kind of work material and the diameter or shape of the hole to be drilled.

The shape and depth of a hole depend greatly on the positional relation between the focus of the laser beam and the surface of the work material. A conical hole is formed when the focus is set below the surface. When the focus is set just on the surface, the hole becomes slightly larger below the surface than at the surface; this tendency increases as the focus position is moved upward from the surface. Machining conditions for deep hole drilling with a YAG laser are shown in Fig. 2.37 [38]. Deep holes have been drilled in a sapphire crystal by ruby laser beam irradiation from opposite surface [39].

Diamond, ruby, sapphire, ceramics, high melting point metals, stainless steel, etc., can be drilled with the YAG laser. It has been reported that the diameter or depth of a hole for various materials can be estimated by an experimentally obtained equation [40]. The depth of the hole is proportional to the laser energy per pulse. A random-walk displacement of hole exit position relative to entrance position, which was caused by multiple reflections at the internal surface of the drilled hole, was reported by Anthony [41]. Special focusing systems for drilling are shown in Fig. 2.38 [42].

The CO_2 laser can successfully drill non-metallic materials, such as glass,

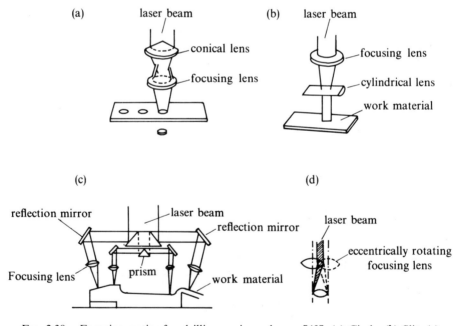

FIG. 2.38. Focusing optics for drilling various shapes [42]. (a) Circle. (b) Slit. (c) Multiple holes. (d) Circle using eccentrically rotating lens

TABLE 2.10
Examples of drilling by CO_2 laser [42]

Work material	Thickness (mm)	Drilling condition	Drilling speed or time
Stainless steel	0.65	1 J, mean 75 W, pulse repeating	10 pulses
Ni alloy	1.78	16 J, pulse duration 0.8 ms	0.8 ms
Alumina	15	Mean 200–400 W, 400 pps[†]	
Alumina	0.7	1 pulse 2 J, 7.5 ms, 0.526 pps	7.5 ms
Alumina	0.63	Mean 75 W, pulse repeating	0.2 s
Natural rubber	0.15	Peak 1 kW, 25 pps, 60 μs	826 mm/min
Fused quartz	5.8	100 W CW	3 s

[†] Pulses per second

TABLE 2.11
Examples of drilling by YAG laser [42]

Work material	Thickness (mm)	Laser power (W) (mean value)	Auxiliary gas	Drilling time (s)
Stainless steel (18–8)	3.05	31	Air	1.33
Ni alloy	1.78	2×10^5 (peak)	–	0.0008
Be	1.24	24	Ar	0.56
W	0.84	31	Air	0.27
Al	1.65	31	Ar	0.28
Ta	3.12	42	Ar	0.36

quartz, ceramics, and plastics, because absorption of a 10.6 μm beam by these materials is very large. Examples of CO_2 laser drilling are shown in Table 2.10, and of YAG laser drilling in Table 2.11.

2.4.2 Cutting and scribing

These processes are among the most widespread applications. There are two cutting methods: one is to cut the workpiece off completely, (often applied for cutting metal sheets); the other is to make a groove on the surface of the workpiece and then to break it (which is applied to brittle inorganic materials). The latter method is called scribing, and CW and high-repetition pulse oscillating modes are used. The latter method of scribing is used in drilling a series of end holes, to avoid thermal fracture.

In Japan, over 70 per cent of the CO_2 laser machining systems installed in factories are used for cutting or scribing metal or non-metal sheets. Only 10 per cent of the applications of YAG laser machining systems are for cutting or scribing. The absorption of CO_2 laser light at the metal surface is not so large as that of the YAG laser, as seen in Fig. 2.2. Therefore, using oxygen as the auxiliary gas, high-speed cutting can be performed, as shown in Fig. 2.39. The auxiliary gas is generally employed to blow off vapour or melt from the groove. In addition to this effect, use of oxygen gas may be expected to increase the cutting speed as a result of the heat of oxidation.

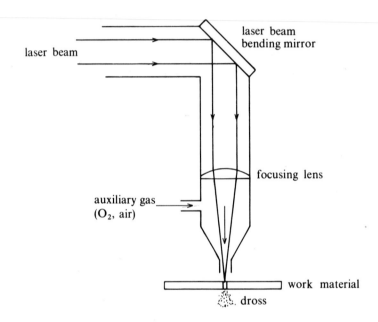

Fig. 2.39. Schematic diagram of laser cutting

Advantages of laser cutting are as follows:

(i) The laser can cut all kinds of materials, regardless of hardness, brittleness, stiffness, etc.
(ii) The kerf width is narrow, and a sharp cut edge can be obtained.
(iii) High speed and precise cutting are possible by selecting a suitable auxiliary gas: N_2, O_2, etc.
(iv) The shape to be cut may be regulated by a numerically controlled work table.

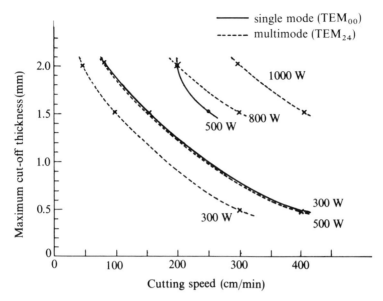

FIG. 2.40. Cutting capability of single- and multi-mode beams. Stainless steel (18–8), thickness 11.5 mm [44]

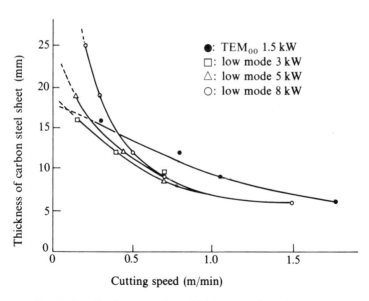

FIG. 2.41. Cutting capability of high-power laser beam [44]

(I) Self burning 75 l/min

(II) Rough-cut 51 l/min

(III) Fine-cut 40 l/min

(IV) Dross 25 l/min

FIG. 2.42. Cutting quality of carbon steel by CO_2 laser [45]. Laser power = 2 kW; Focal length of lens = 127 mm (ZnSe); cutting speed = 1 m/min, thickness = 5 mm

(v) Distortion of the cut parts is very small.
(vi) Production process planning, change of shape of products, and nesting layout are made very easy by computer control.
(vii) Reliable and high-quality cutting in unmanned operation can be expected.

The quality of the cut surface is evaluated mainly by the quantity of dross remaining, the roughness of the cut surface, and the kerf width. The relations between cut thickness of a stainless steel plate and laser power and beam mode are shown in Fig. 2.40 [43]. The speed of cutting of thick sheet metals is determined more by the beam mode than by the power of the laser. The cutting ability of a single-mode, high-power beam is shown in Fig. 2.41 [44].

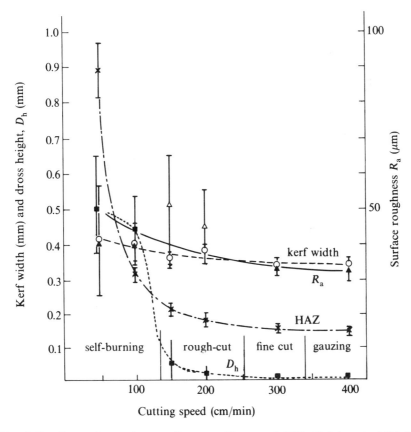

FIG. 2.43. Dependence of cut quality on cutting speed [45]. Stainless steel (18–8); thickness = 1.5 mm; laser power = 1000 W; focal length = 62.5 mm; power density = 3.9×10^5 W/cm^2; spot diameter = 0.48 mm; HAZ = width of heat-affected zone; R_a = surface roughness; D_h = dross height

The limit of thickness that can be cut off by a 8 kW laser at 0.2 m/min cutting speed is about 25 mm.

Cutting may be classified into gauzing (not shown in the figure), dross, fine cut, rough cut, and self-burning, according to the cross-sectional form of the kerf as seen in Fig. 2.42 [45]. Not only kerf form or width but also roughness of cut surface, heat-affected zone width, and quantity of dross are affected by cutting speed, as is shown in Fig. 2.43. The periodic nature of the cut surface roughness at low cutting speed is explained by the relation between the speed of motion of the beam and the rate of propagation of the oxidizing reaction.

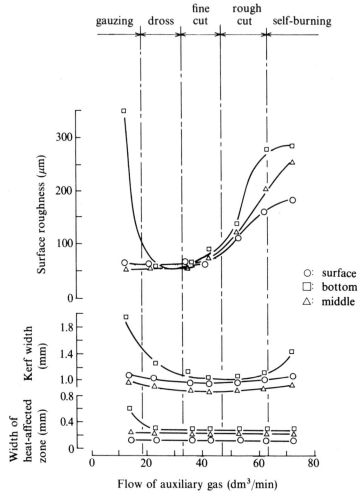

FIG. 2.44. Relation between cut quality and flow rate of auxiliary gas [45]

The effect of the auxiliary oxygen gas on cutting quality is summarized in Fig. 2.44. The positional relation between the focal point of the beam and the surface of the work material affects the quality of the cut surface (Fig. 2.45). From Figs. 2.44 and 2.45, it is seen that the kerf is narrow and the roughness of cut surfaces is fairly small. The cutting speed v (mm/min) for metal sheets can be estimated by the following experimentally derived relation:

$$v = k \frac{Q}{L_v d l} \qquad (2.18)$$

where Q is the laser power (W), L_v is the latent heat of vaporization (J/kg), d is the spot diameter (m), l is the thickness of the sheet (m), and k is an arbitrary constant.

The cutting capabilities of three cutting methods—laser, gas, and plasma—are shown for a stainless and a carbon steel in Fig. 2.46 [46]. If oxygen is used

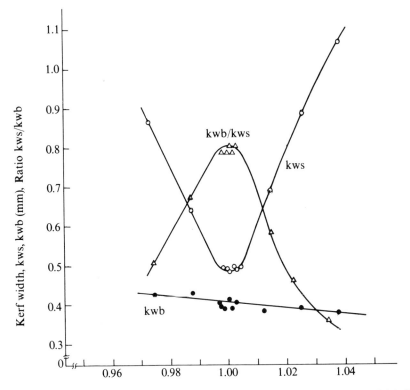

FIG. 2.45. Effect of beam waist on quality of cut surface [44]. Stainless steel (18–8); thickness = 1.5 mm; power = 1200 W; cutting speed = 300 mm/min; focal length = 125 mm; kws = surface kerf width; kwb = bottom kerf width

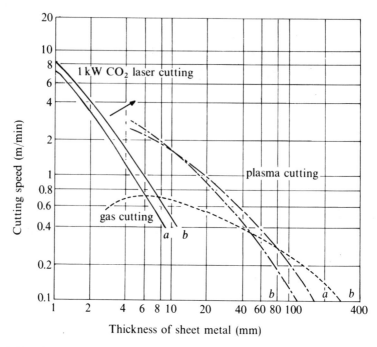

Fig. 2.46. Comparison of different cutting methods. (a) Stainless steel. (b) Structural steel [46]

as auxiliary gas, the cutting speed for steel tends to increase, so the graph will move in the direction denoted by the arrow in the figure. Results obtained for cutting several representative materials are listed in Tables 2.12, 2.13, and 2.14. Composite materials, such as epoxy resin sheet reinforced with glass or

TABLE 2.12
Non-metal cutting with CO_2 laser [30]

Material	Thickness (mm)	Speed (mm/min)	Kerf width (mm)	Power (kW)	Auxiliary gas
Glass	9.53	1520	1.0	20	
Fused quartz	9.5	130		1	
Quartz	1.9	600	0.2	0.3	
Concrete	38	50	6.35	8	Air
Ceramic tile	6.4	500		0.85	
Limestone	38	130		3.5	

TABLE 2.13
Steel cutting [30]

Material	Thickness (mm)	Speed (mm/min)	Kerf width (mm)	Power (kW)
Mild steel	0.3	6000	0.1	0.3
	1	4500	0.05–0.1	0.4
	2	4900	0.1–0.2	2
	3.18	890	0.5	0.4
	6.4	2300	1.0	15
Stainless steel (18–8)	0.45	634	0.46–0.92	0.2
	1.27	760	0.51	0.165
	2.0	1800		1
	3.18	5080	0.1–0.2	0.5
	4.75	1270	2.0	20
Tool steel	3.0	1700	0.2	0.4
High-speed steel	2.3	1100	–	0.5
High-tension steel	3.2	760	–	0.25

TABLE 2.14
Metal cutting [30]

Material	Thickness (mm)	Speed (mm/min)	Kerf width (mm)	Power (kW)
Ti	0.508	203	0.46–0.92	0.2
	1.52	15 200	0.4	0.135
	5.0	3 300	0.4	0.85
	10	2 526	1.7	0.25
	19.05	2 540	0.1–0.05	10
Ti alloy	1.3	7 600	0.8	0.21
	5.0	4 000	0.4	1.0
	10	2 500	1.6	0.26
Al	1.5	2 500	0.8	1.0
	12.7	2 300	1.0	15
Inconel 718	12.7	1 270	–	16

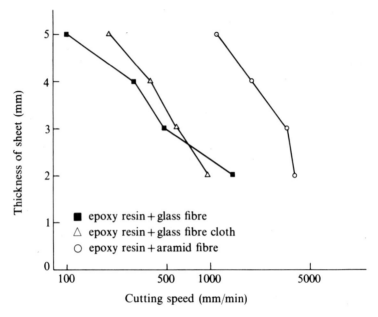

FIG. 2.47. Cutting of fibre-reinforced material. Laser power = 400 W; single mode [47]

polyamide fibre have been cut excellently (Fig. 2.47) [47] and this technology has been applied to production processes in the aircraft industry.

2.4.3 Micro-machining [48, 49]

For this process the YAG laser oscillator is generally employed, but sometimes CO_2, Ar ion, or Nd–glass laser oscillators are used. These processes are applied mainly in making electronic parts. The relation between oscillating mode (CW, pulse) and type of application is given in Table 2.15. Practical applications include trimming, marking, balancing, tuning, and repair. Machining systems controlled by a micro- or a mini-computer have been developed for various applications.

Laser-beam trimming is used to adjust the resistance or capacitance of hybrid integrated circuits by removing part of the pattern formed on alumina ceramic substrates. This method is excellent, comparing favourably with sand blasting or anodic oxidation in the following features:

(i) precise adjustment of removed volume
(ii) high-speed control of beam, size, power, and position
(iii) adjustment during operating condition.

TABLE 2.15
Oscillating modes used for micro-machining [49]

Excitation mode	Oscillating mode	Applications
Continuous	Continuous	Cutting
	Q-switch, high repetition	Trimming, scribing, marking, drilling, tuning
Pulse	Pulse	Scribing, marking, repair
	Normal	Cutting, marking, drilling
	Q-switch	Repairing, balancing

Laser-beam marking is used to engrave model number, manufacturer's serial number, manufacturing date, lot number, etc., on products. There are two marking methods. Marks or characters in a metal mask (which is cut with a laser beam) are projected and engraved at reduced size on the surface of parts by a magnified laser beam. This method is convenient for marking a large number of items with the same pattern at one time. When many different marks are required, or when the batch size is small, the computer-controlled beam-positioning and scanning system shown in Fig. 2.30 is used. Semiconductor pellets are laser-marked instead of ink-stamping or manual writing with a diamond pen in many industries, because of adaptability to any production process, high quality of marks, high reliability, and few obstacles to products. Laser marking has been a useful tool in quality control systems.

Recently, the possibility has been considered of applying laser marking in production process control. For instance, in the automobile industry it is planned to read automatically marks engraved on the body frame, and to distinguish the kind of car in order to control the assembly process. In this case, marks carry not only quality information but also assembly information. These applications of marking will increase.

In balancing of rotating components, the position and amount of imbalance are measured by rotation, and the unbalanced weight is removed by laser pulse irradiation during rotation. Using laser beam, it is possible to detect and to regulate imbalance during rotation. The imbalance regulation of a revolving body, such as a gyroscope, rotating at high speed (10 000 to 25 000 rev/min), using a laser beam reduces the processing time from 3–5 h to 10–15 min. This process has been applied practically to the rotor of a motor, the steam turbine, a spinning rotor, etc.

Laser-beam tuning is the process of finely adjusting the oscillating frequency of a quartz oscillator during operation, by removing an extremely small volume with a laser pulse.

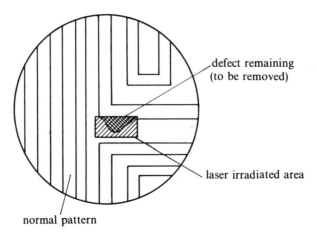

Fig. 2.48. Schematic diagram of mask repair

Laser-beam repair is the process of correcting residual defects on a photomask for IC or LSI by direct removal with a pulsed laser beam focused into a rectangular shape, as shown in Fig. 2.48.

Applications of machining without contact with the work material, which is one of the excellent features of laser processing, will gradually increase in many production processes.

2.5 Welding and soldering

2.5.1 Welding of metals

The high-power CO_2 laser beam may be used effectively for deep penetration welding with a narrow fusion zone under atmospheric conditions. Energy efficiency, which is expressed as the ratio of energy actually needed for welding (estimated from fused volume) to input energy, is very high compared with that of other welding methods except electron-beam welding. Moreover, the heat-affected zone is narrow, and the residual stress or strain is also small. Figure 2.49 shows the energy per unit area necessary for welding by several methods [50].

The relation between penetration depth and welding speed in laser-beam welding is compared with that for electron-beam welding in Fig. 2.50 [51]. From these figures it is seen that laser-beam welding of sheet metal less than 15 mm in thickness is almost equivalent to electron-beam welding in terms of penetration depth and energy requirement.

Penetration depth is influenced not only by beam mode but also by the existence of plasma fume (superheated metal vapour) in the beam path. The

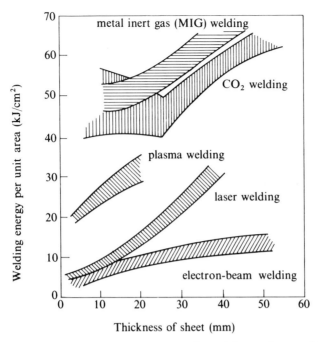

Fig. 2.49. Energy per unit area necessary for welding carbon steel [50].

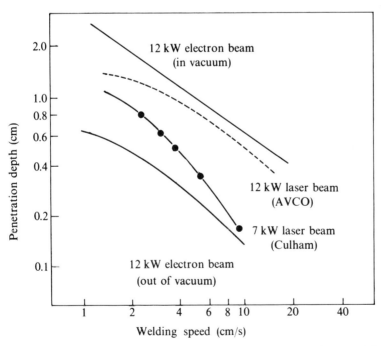

Fig. 2.50. Comparison of penetration depth in laser and electron-beam welding of carbon steel [51]

metal vapour is generated during welding and superheated by absorbing laser power. The vapour is rapidly heated to an extremely high temperature and enters the plasma state. The maximum penetration depth is obtained by a focused single-mode (Gaussian mode) beam, but in practice it is very difficult to oscillate an ideal single-mode beam at high power level. Furthermore, it is very important to select the most suitable focusing systems so as to obtain the highest energy density of the beam. Deep penetration is obtained by a beam of energy density above a threshold value. The energy efficiency in welding decreases with increasing input laser power. Penetration depth is increased by blowing away the plasma fume with a gas jet from the side [52]. The effect of the gas jet depends on the kind of gas available and the blowing direction. A melt of a good cross-sectional shape and high stability is obtained with a gas

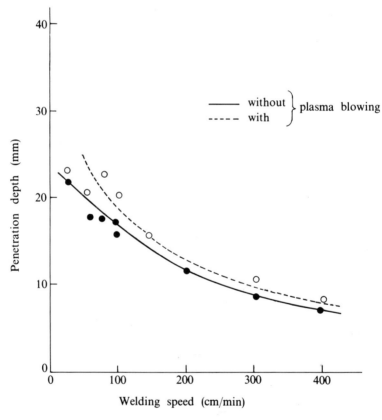

FIG. 2.51. Effect of blowing away of plasma on penetration depth [43]. Work material = carbon steel; laser power = 1.5 kW; focal length = 127 mm; spot diameter = 0.45 mm; gas shield = Ar; gas jet = Ar + He

jet blown parallel to the work surface. The effect of a gas jet on penetration is shown in Fig. 2.51.

The penetration depth or cross-sectional shape of a weld is apparently influenced by the position of the focal point of the laser beam [53]. Figure 2.52 shows the effects of focal position on the depth and width of the weld. The horizontal axis denotes the ratio (which is called the A_b value) of the distance between lens and work surface to the focal length. The maximum penetration is obtained with the focal point positioned slightly below the surface: 0.97 or 0.98 of the A_b value. This is explained by the fact that the true focusing length becomes shorter than the focal length of the projecting lens as a result of spherical aberration and change in focal length due to thermal expansion of the lens caused by absorbed laser energy.

Penetration depth increases in proportion to laser power (Fig. 2.53) [54]. Materials of low thermal conductivity and high melting point are more susceptible to the effect of welding speed.

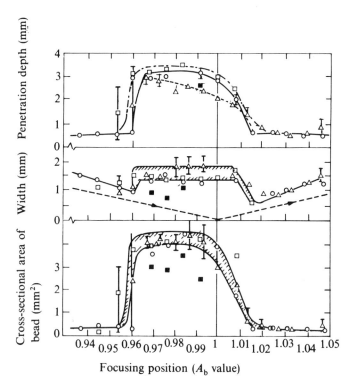

FIG. 2.52. Relation between focusing position and weld bead quality [53]. Work material = stainless steel (18–8); laser power = 1 kW; welding speed = 0.5 m/min; focal length = △–38, ○–64, □–127, ■–254 mm

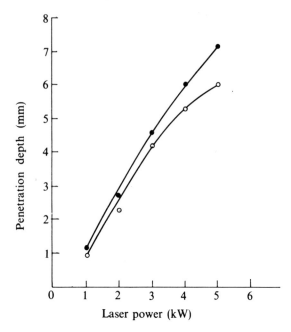

FIG. 2.53. Relation between penetration depth and CO_2 laser power [54]. Work material = stainless steel (18–8); focal length = 127 mm; welding speed = ●–1, ○–1.8 m/min; shield gas = Ar, 15 dm^3/min; blowing-off gas jet = He, 50 dm^3/min

Laser welding is applied to many kinds of metals (Table 2.16) and employed in various industries (Table 2.17). One application of laser welding is the joining of hot coils in continuous pickling line using a CO_2 5 kW laser and a filler-wire supply device [55].

2.5.2 Welding of ceramics [56]

It is possible to weld ceramics only with a laser beam; it seems to be impossible by any other method. Ceramics are refractory, hard, wear-resistant, insulating materials, used in various products. Welding technology using a laser beam is expected to create new fields of application for ceramics. When ceramics are drilled, cut, or welded by a laser beam, the big problem is the frequent occurrence of cracks during or after irradiation, and sometimes the workpiece is broken. To avoid generation of cracks, a pulsed laser is generally used for drilling and cutting. Microcracks induced in the material reduce the strength of the welded workpiece.

To prevent generation of cracks, workpieces are preheated before welding and cooled gradually in a furnace. Suitable rates of heating and cooling

TABLE 2.16
Welding of various metals [30]

Metal		Thickness (mm)	Welding speed (cm/s)	Laser power (kW)	Spot size (mm)
Thin sheet	Stainless steel	0.127	2.12	0.25	0.2
	Tinplate	0.178	1.91	0.25	0.2
	Ti	0.25	12.7	0.25	0.2
	Tic–P	0.127	5.92	0.25	0.2
	Ni	0.127	1.48	0.25	0.2
	Inconel 600	0.102	6.38	0.25	0.2
	Carbon steel	0.79	0.85	0.6	
Thick plate	High-strength steel	12.0	1.27	10.6	
	Al alloy	12.0	2.75	8.0	
	Ti alloy	12.0	1.48	11.0	
	Stainless steel	6.5	0.6	2.0	
	Mild steel	4.0	0.7	2.0	
	16% Cr 10%Ni alloy	4.75	2.5	2.0	

TABLE 2.17
Examples of CO_2 laser welding [42]

Laser power (kW)	Applications
0.05	Miniature battery cells
0.25	Honeycombs, bellows, cases for electronic equipment
0.35	Coffee percolators
0.45	Ball-pen cartridges, valves, diaphragms
0.5	Ti vessels, gyrocompasses, bearings, plastic-covered wire
1	Yokes of automobile heatmotors roller bearings, band saws, silicon-steel sheet
2	Timing gear, high-pressure vessels, Al vessels
5	Steel plate for rolling
8	Transmission gear, ship bodies

depend on the size and kind of material to be welded. The preheating temperature should be determined according to the melting point. Work materials are usually heated to one-third or one-half of the melting temperature. An electric or gas furnace is used for preheating and gradual cooling.

Figure 2.54 shows cross-sectional views of butt-welded ceramics. These ceramics are welded well and no microcracks are observed. The photograph on the left shows welding of two pieces of alumina, but it is possible to weld two different kinds of ceramics with quite different thermal and mechanical properties (Table 2.18), as shown on the right of Fig. 2.54. Workpieces are heated to about 1100 K before welding, and cooled to about 600 K for 5 min after welding. The welded penetration depth, D (mm) increases with decreasing welding speed, V (mm/min). The relation $D = AV^k$ has been derived experimentally, where A and k are constants that depend on the kind of material. In practice, when two workpieces are set with a gap of about 50 μm, deeper penetration depth can be obtained. Because the laser beam penetrates into the gap, and both walls facing into the gap are melted, the melted region is deep. In this case the surface of the melt becomes concave, because the melt fills the gap.

In the bending test, the strength of a butt-welded joint is about 70–75 per cent of that of the original materials, under optimum welding conditions. This decrease in strength is considered to result from the generation of microcracks and porosity, or from crystal growth in the melt region. If the temperature gradients for preheating and for subsequent cooling are too large, large thermal stresses arise and generate microcracks. Pores existing originally in

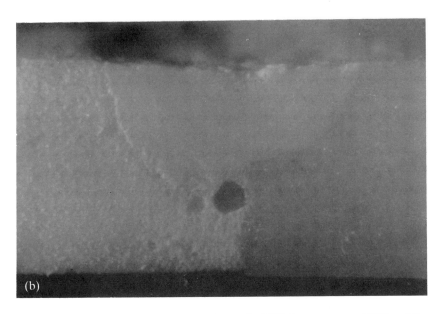

Fig. 2.54. Cross-sections of butt-welded ceramics [56]. Laser power = 80 W; welding speed = 5 mm/s; thickness of workpiece = 1 mm; preheating to 1100 K. (a) Alumina–alumina. (b) Forsterite–alumina

TABLE 2.18
Thermal and mechanical properties of forsterite and alumina [57]

	Forsterite	Alumina
Bulk density (kg/m^3)	2880	3400–3700
Melting point (K)	2163	2373
Thermal expansion coefficient 300–10000 (10^{-6}/K)	11.2	7.5–7.9
Thermal conductivity (W/m K)	3.36	16.8–18.9
Tensile strength (MPa)	67.2	13.5–16.8
Compressive strength (MPa)	57.7	96.1–269
Bending strength (MPa)	13.5	26.9–40.4
Impact strength (kJ/m^2)	34.3	54.9–60.8
Water absorption (%)	0–0.002	0–0.00
Firing temperature (K)	2163	2073

the melt area show a tendency to migrate towards the periphery of the melt, as seen in the photograph on the right in Fig 2.54. To control the initiation of microcracks, it is necessary to cool the specimen gradually. The size of crystals within the melt region becomes larger than that of the original matrix during the gradual cooling process. This crystal growth results in a reduction in strength. It seems to be impossible to obtain the same strength as that of the original matrix by welding.

Research has been directed at welding ceramics with neither preheating nor gradual cooling. In Fig. 2.55, samples of alumina plates welded by a Gaussian-mode laser beam are shown. The welding speed is very slow and the bead width is large. The furnace method of preheating and gradual cooling is not employed, but welding is carried out by a defocused beam. Laser energy at the periphery of a defocused beam, where the power density is not so high, is used for preheating and gradual cooling, and the energy at the centre of the beam, where the power density is sufficiently high, is used for welding. Using this technique, it is possible to weld ceramics in the same way as metals, without any other special heat treatment [57].

Alumina pipes welded using two lasers are shown in Fig. 2.56 [58]. One laser, with programmed output power, is used for preheating and gradual cooling. Another is used for welding. The pipes to be welded are fastened to a chuck and rotated at 90 rev/min. Then a laser beam, whose power increases at the programmed rate, is directed onto the joint between the pipes. When the temperature rises to about 1100 K, rotation is reduced to 2 rev/min and the laser beam is cut off. Another laser then begins to irradiate the pipes for welding. Welding is completed within one revolution, and then the rotation

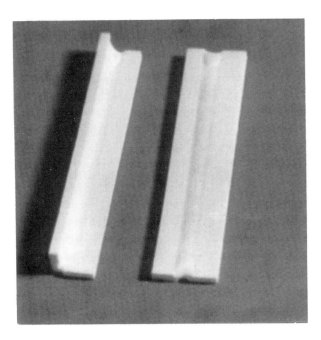

FIG. 2.55. Alumina ceramic plate welding, using neither preheating nor gradual cooling [57]. Laser = CO_2 (transverse mode TEM_{00}); laser power = 140 W; welding speed = 24 mm/min; thickness of workpiece = 2 mm

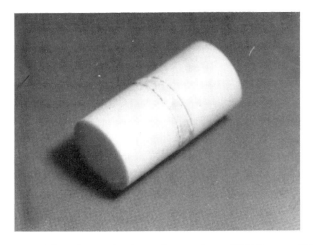

FIG. 2.56. Butt welding of ceramic pipes using two lasers [58]. Laser = CO_2 (TEM_{00}); laser power = 210 W; welding speed = 30 mm/min; outside diameter = 20 mm; inside diameter = 16 mm

speed is increased to 90 rev/min. The programmed laser beam is then projected onto the workpiece again to allow it to cool gradually to about 600 K.

2.5.3 Micro-welding and soldering

The YAG laser is generally used for micro-welding and soldering, because for these processes, a power density below 10^4 W/cm^2 is necessary, and the power density of the YAG laser can be controlled by defocusing. When welding is performed by a beam of limited power density, with which no splash is observed, maximum bonding strength can be obtained. Supposing that the power density distribution is Gaussian, the power density threshold Q_v(W/cm^2) at which splash does not occur is expressed theoretically by the following formula [59]:

$$Q_v = \frac{T'_v \lambda \pi^{\frac{1}{2}}}{a \tan^{-1}\left(\frac{4\kappa t}{a^2}\right)^{\frac{1}{2}}} \qquad (2.19)$$

where λ is the thermal conductivity, a is the radius of the heated area, t is the pulse duration, and κ is the thermal diffusivity. Corrected boiling point, T'_v is derived from the boiling point, T_v, latent heat of fusion, L_m, and latent heat of vaporization, L_v, by the following equation:

$$T'_v = T_v + \frac{L_m + L_v}{c} \qquad (2.20)$$

where c is the specific heat. Q_v values determined experimentally for several metals agree well with the calculated values (Fig. 2.57) [60].

Micro-welding is applied in the manufacture of cathodes for Braun tubes or CRTs [60], relay contact points [61], metal casing for semiconductor devices, magnetic heads [62], etc.

Recently, soldering with lasers has been introduced [63]. The following features are especially expected in laser soldering:

(i) It is possible to solder an extremely narrow area of the order of 10^1 μm.
(ii) There is little thermal effect or thermal damage to the circumferential area.
(iii) Protuberances do not occur and reliability is very high.
(iv) Programme-controlled operation is possible.

2.6 Surface treatment

Surface treatment technologies using a laser beam may be grouped as shown in Fig. 2.58 on the basis of Gnanamuthu's classification [64].

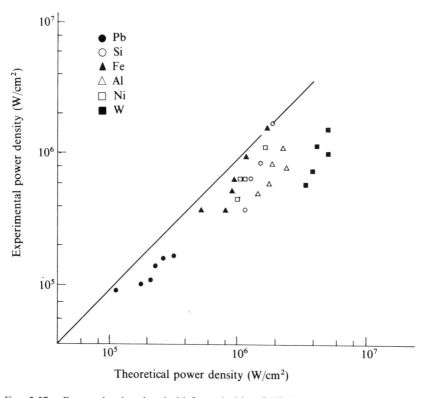

Fig. 2.57. Power density threshold for splashing [60]. Laser energy < 10 J; pulse duration = 0.2 ~ 8 ms; focal length = 50 mm

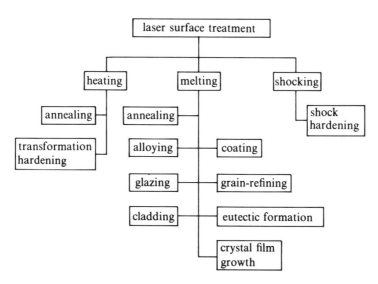

Fig. 2.58. Current methods of laser surface treatment of materials [63]

2.6.1 Hardening and annealing

Surface hardening is applied more in practice than other laser surface-treating technologies. When a carbon steel surface is heated rapidly by the laser beam, the irradiated portion of the workpiece surface is locally heated above the transformation temperature. Then the carbon in the steel diffuses into an austenite matrix. After irradiation is stopped, the austenite is not transformed to pearlite, but is converted to martensite by rapid self-cooling as a result of heat flow into the surrounding material. This process is called laser hardening.

Features of laser hardening are:

(i) only the surface layer or a limited small area can be hardened;
(ii) thermal deformation does not occur;
(iii) the inside wall of a pipe or side wall of a narrow groove can be hardened easily;
(iv) treatment can be effected quickly on a production line.

The mode of the applied laser beam directly affects the results of hardening. Temperature distributions of several beam modes, such as the Gaussian, rectangular, annular (see Fig. 2.32), and semicircular, have been calculated. Calculated temperature distributions in the workpiece as a result of irradiation by the annular-mode laser are shown in Fig. 2.59. This mode is more suitable than the Gaussian for hardening [65]. There is no large difference between the surface temperatures calculated by one- and three-dimensional

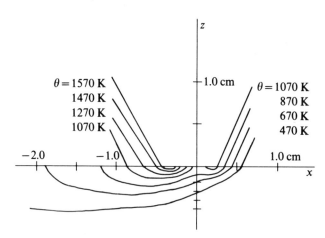

FIG. 2.59. Temperature distribution in a carbon steel workpiece for annular laser mode [65]. Laser power = 2 kW; beam diameter ≈ 9.5 mm; travel speed = 45 cm/min; distance between focusing point and workpiece = 70 mm

TABLE 2.19
Experimental results for hardening of cast irons [65]

Type of iron	Beam scanning velocity V (cm/min)	Hardened depth h (mm)	Hardened width W (mm)	Vickers hardness H_v
Grey cast iron (FC30)	250	0.32	2.70	744
Nodular iron (FCMP)	250	0.44	3.38	754
Ductile iron	100	0.46	3.18	708
Carbon steel (45C)	250	0.60	3.98	705

Laser power = 2 kW; mode = TEMoo; spot size = 7.3 mm (5.0 mm for 45 C); coating = $Mn_3(PO_4)_2 \cdot 3H_2O$

heat flow models at beam scanning speeds of over 2 m/min. Some experimental results are given in Table 2.19. Exhaust valves, valve seats, cylinder bores, cam and crank shafts, gear teeth, piston ring grooves, and automobile steering gear housings may be laser-hardened. Some of these are current uses; others are under development. Such applications will increase rapidly in the near future.

Various kinds of laser annealing processes are becoming available. The most promising application is the annealing of semiconductor crystals [66]. Partial annealing of metals is also used. Generally speaking, laser annealing is the process used to recover the atomic arrangement in semiconductor crystals disordered during impurity-doping by ion implantation or thermal diffusion. Laser annealing does not bring about any change in impurity distribution or contamination with other impurities. In the annealing process using short-pulse laser irradiation, a thin surface layer less than a few hundred nanometres thick is melted and recrystallization occurs within 1 μs. In the process using a CW laser, the surface does not melt and recrystallization occurs within a few microseconds. From these technologies it is expected that a technique will be developed for forming a single-crystal film.

2.6.2 Alloying [67]

This process is used for melting part of an iron surface previously coated with temper metal powder, to form an alloy layer [68]. Temper metals include Cr, Ni, V, and Co. Experimental results are shown in Fig. 2.60. Alloyed layers of about 1.5 mm thickness are obtained. According to X-ray analysis and electron probe microanalysis (EPMA), homogeneous temper metal diffusions and alloy structures are observed. Few applications have yet been introduced

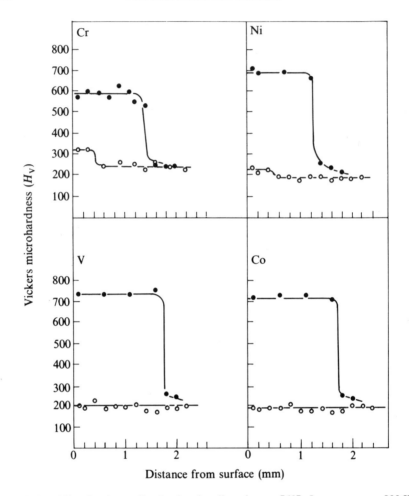

FIG. 2.60. Microhardness distribution in alloyed area [68]. Laser power = 800 W (CO_2); travel speed = 50 cm/min; focal length of lens = 95 mm (ZnSe)

into industry. In the near future there will be an increase in practical uses for conserving resources and keeping production costs down.

Glazing [69] is a new process for producing rapidly chilled metallurgical microstructure or glassy structure surface layers at cooling rates of over 5×10^6 K/s, by high-speed scanning of a laser beam of high power density. These amorphous or microstructure films are formed with coated metals. The films show corrosion-resistance or wear-resistance. The technology is at present only experimental.

2.7 New applications

Applications of the laser beam as an effective and controllable energy source have been studied. Many experimental results have been presented. These studies will expand and create new practical application fields.

2.7.1 Laser-assisted cutting [70]

This method is used to machine hard materials such as Inconel 718 or titanium, because these materials can be weakened or softened by heating with a laser beam, before or during cutting (Fig. 2.61). When Inconel irradiated by a CW CO_2 laser beam is cut, the effects of reducing the cutting force, such as prolonging the ceramic cutting tool life and increasing the metal removal rate, are confirmed experimentally. Also, Ti 6–4 bars, previously irradiated by a pulse laser beam, may be cut with lower cutting force. These results suggest the possibility of cutting hard materials in the same way as softer metals witha a lathe.

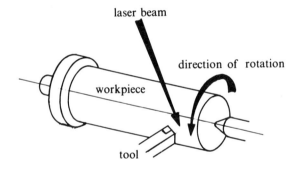

FIG. 2.61. Schematic diagram of laser-assisted turning [70]

2.7.2 Laser coating and production of ceramic–metal eutectic [71]

Laser coating is a new kind of surface treatment technology. A ceramic coating technique has been developed using a high-power CO_2 laser as a heat source for evaporating ceramic materials. The ceramic films deposited are very hard and firm, and adhesion to the substrate is also high. A diagram of the experimental set-up is shown in Fig. 2.62. Target rings of 30 mm in outer diameter were preheated to 1100 K by an electric heater and rotated slowly to avoid thermal fracture. A molybdenum substrate was also preheated to 600–900 K. The targets were irradiated by the laser beam, the power of which was gradually increased to 500 W. The power density was selected

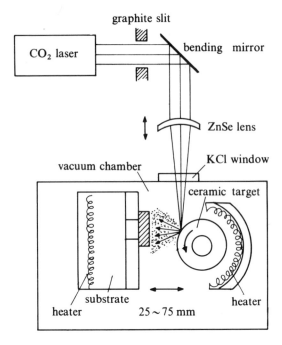

FIG. 2.62. Schematic diagram of ceramic coating apparatus [71]

as $1\text{--}2.5 \times 10^4$ W/cm^2. The composition and Knoop hardness of the target materials, and the Knoop hardness and deposition rate of the coated film are summarized in Table 2.20. It should be noted that the hardness of a coated BN(1) film is extremely high, notwithstanding the very low hardness of the initial hexagonal-structure target material.

SEM photographs of forcibly fractured mullite and BN(1) films are shown in Fig. 2.63. On the broken sections of both films there are ripple marks which suggest that the films are hard, brittle, and firm. The adhesion of films to the substrates is so strong as to be not measurable.

An experimental study of laser-enhanced gold plating has been reported. A new technique has been studied for making metal–ceramic eutectic surfaces on aluminium substrates by coating with Al, Ti, and Cr powders and laser irradiation in gases of various compositions [72]. The wear characteristics of these surfaces have also been reported.

2.7.3 Laser shaping

Three-dimensional shapes of ceramics are cut efficiently by a high-power CO$_2$ laser beam focused with special optics to form a kind of a milling machine

TABLE 2.20
Characteristics of target ceramics and coating films [71]

Target ceramics	Composition (wt%)	Knoop hardness	Coating film	
			Deposition rate (μm/min) (at 320 W laser power)	Knoop hardness
Al_2O_3	Al_2O_3 99%	1568	0.65	2166
$3Al_2O_3$–$2SiO_2$	Al_2O_3 73%, SiO_2 26%	451	0.28	1235
Si_3N_4	Post-sintering 20%, 80%	588	0.35	1068
Sialon	Si_3N_4 76.8%, Al_2O_3 23%	1568	0.18	1164
BN (1)	HBN 95% Binder CaO, B_2O_3	53	0.06	4225
BN (2)	HBN 60% Al_2O_3 20%, SiO_2 20%	36		2518

HBN = hexagonal boron nitride

film ←/→ substrate |—10 μm—|

FIG. 2.63. SEM photographs of fractured coating ceramic films [71]. (a) Mullite. (b) Cubic BN. Laser power = 40 ~ 400 W (CO_2); irradiation time = 5 ~ 20 min; focal length of lens = 300 mm (ZnSe); pressure < 10^{-4} torr; distance between target and specimen = 20 ~ 50 mm

[73]. A screw thread was machined experimentally on rods of SiC, Si_3N_4, Sialon, and alumina. The probability of fracture in laser-machined samples was compared with that in diamond-machined ones, and it was found that the fracture strength of the laser-machined samples was lower but that the range of strength was extremely narrow compared with the diamond-machined samples. For a turbine blade of hot-pressed silicon nitride, the cost of cutting by laser milling is estimated as one-third of that by diamond machining.

2.7.4 Laser-enhanced etching and plating

A laser beam can be used to initiate or to enhance etching processes. An etched pattern can be developed on the laser-irradiated portion of the workpiece [74]. This will be a promising micro-fabrication process in the near future. This kind of pyrolytic etch process is based on the thermal effect which enhances etching on the laser-irradiated area.

Etching due to the photocatalytic effect of laser irradiation is called laser photolysis.

A laser-enhanced etching process in an etch gas has been reported for single-crystal or polycrystalline Si. Micro-fabrication experiments were carried out on an Si crystal in a gaseous environment of Cl_2, or KCl and SF_4, with a CO_2 or argon–ion laser [75].

 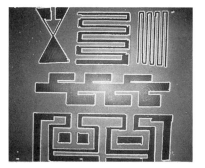

FIG. 2.64. Exterior view of laser-enhanced etching in KOH solution (alumina–TiC) [74]. Laser power = 1 W (20 kHz pulse YAG); beam mode = Gaussian; travel speed = 4 mm/min; concentration of solution = 40%

Alumina–TiC etching was enhanced with an Ar ion or YAG laser in potassium hydroxide solution [76]. Figure 2.64 shows an example of alumina–TiC grooves made by laser-enhanced etching.

Laser-enhanced electro-plating can be applied for patternmaking using a krypton laser. Heatless etching of biological and polymeric materials has also been developed at IBM using a far-ultra-violet laser beam [77].

2.7.5 Crystal growth and other applications

The possibility of forming single-crystal films from the liquid phase by laser-beam irradiation was described in the section on laser annealing. Experimental results have been reported for lateral epitaxial crystal growth of silicon on areas of silicon dioxide with a CW Ar ion laser. This process is considered to be one of key technologies applicable to VLSI production. Large-grained macrocrystalline silicon in ribbon form was drawn at a high rate by a pair of scanned focused CO_2 laser beams [78]. Figure 2.65 is a photograph of sapphire crystal lands grown instantaneously by focused CW CO_2 laser-beam scanning. Fibre crystal growth was also observed [79].

Laser energy is used as a heat source for various applications as follows: optical fibre-drawing [80], generation of extremely clean surfaces [81], removal of selective defects from a slab [82], control of generation of thermal fracture of a rock [83], etc.

2.8 Conclusions

Processing of materials is an important application of lasers. This technology has been applied practically in many manufacturing industries. To extend

FIG. 2.65. SEM photograph of sapphire crystal lands grown on sapphire substrate [79]. Laser power = 100 W (CO_2); spot size = 0.2 mm; travel speed = 5 mm/s

laser processing to wider fields, it is essential to develop a laser processing machine, with better characteristics of stability of output power, stability of beam mode, stability of focus positioning, reduction of initial and operating costs, safety systems for the operator, etc. Furthermore, laser-beam machining has so far been introduced in many factories only on account of its high efficiency or high machining speed. However, new applications should be developed, making use of special characteristics of the process which cannot be achieved by other machining processes.

Laser-beam processing is highly flexible as shown by the following:

(i) Various types of machining, such as cutting, welding, drilling, and hardening, can be done with a laser beam by controlling oscillation and beam mode, that is, output power, power density, irradiation time, and so on.

(ii) Without changing tools or machines, various kinds of work material can be machined effectively, including ceramics and composite materials.

(iii) It is very easy to transfer a laser beam to several work stations sequentially by bending and changing the optical path. The laser beam can be guided into freely selected gas atmospheres, including a vacuum.

(iv) No machining force acts on the workpiece during machining. It is not necessary to fix the workpiece firmly on the jig.

(v) Laser processing can be carried out in-process, together with measurement or other machining techniques.

A new flexible manufacturing system using a laser (LFMS) which can fabricate various kinds of products in small batches is currently being developed.

A guarantee of safety in laser operation is very important. Technical Committee No. 76 of the International Electrotechnical Commission submitted a document, 'Radiation safety of laser products, equipment classification, requirement, and user's guide' in July 1982. Every country is currently enforcing or establishing standards, rules, or guides for laser safety based on this document. Lasers used for processing are grouped in Class IV. Stipulation of safety precautions, hazards incidental to laser operation, procedures for hazard control, and maximum permissible radiation exposures is required, and laser equipment makers and operators are placed under an obligation in the document.

References

[1] Maiman, T. H., *Phys. Rev. Letters* **4** (1960), 561.
———, *Nature* **187** (1960), 493.
[2] Javan, A. *et al.*, *Phys. Rev. Letters* **6** (1961), 106.
[3] Snitzer, E., *Phys. Rev. Letters* **7** (1961), 444.
[4] Hall, R. N. *et al.*, *Phys. Rev. Letters* **9** (1962), 366.
Nathan, M. I. *et al. Appl. Phys. Letters* **1** (1962), 62.
[5] Bridge, W. B., *Appl. Phys. Letters* **4** (1964), 128.
——— and Chester, A. N., *Appl. Optics* **4** (1965), 573.
[6] Geusic, J. E. *et al.*, *Appl. Phys. Letters* **4** (1964), 182.
——— ibid. **6** (1965), 175.
[7] Patel, C. K. N., *Phys. Rev. Letters* **12** (1964), 588.
———, ibid. **13** (1964), 617.
———, *Phys. Rev.* **136** (1964), A1187.
[8] Schafer, F. P. *et al.*, *Appl. Phys. Letters* **9** (1966), 306.
Sorokin, P. P. and Lankard, J. R., *IBM J. Res. Develop.* **10** (1966), 162.
[9] Charschan, S. S. and Barros, C. S., *Laser Focus* **18** (1982), 24.
Epperson, J. P. *et al.*, *Western Electric Engineer* **10** (1969), 2.
[10] Adachi, K. (Editor), *Electronic theory of metals* (Maruzen, 1973), 275 (Japanese).
[11] Cohen, M. I. and Epperson, J. P., *Electron-beam and laser-beam technology* (Academic Press, 1968), 118.
[12] Washburn, E. W. editor, *Intern. Crit. Table of Numerical Data, Phys. Chem. & Tech.* **V** (McGraw-Hill, 1929), 248.
[13] Ikeda, M. *et al. Preprint of Meeting for Interim Report of National R & D Program 'FMSC provided with laser'* (1982), 47 (Japanese).
[14] Tsukamoto, K. *et al.*, *Proceedings of the 3rd ICPE* (1977), 155.
[15] Pittaway, L. G., *Brit. J. Appl. Phys.* **15** (1964), 967.

[16] Carslaw, H. S. and Jaeger, J. C., *Conduction of heat in solids* (OUP, NY, 1959), 264.
[17] Cohen, M. I., *Laser handbook*, vol. 2, Arecchi, F. T. and Schulz-Dubis, E. O., editors (North-Holland, 1972), 1577.
[18] Namba, S. et al., *Laser processing* (Nikkankogyoshinbun-shya, 1972), 92 (Japanese).
[19] Miyazaki, T. and Taniguchi, N., *J. Japan Soc. of Precision Engg.* **36** (1970), 21 (Japanese).
[20] Ikeda, M. et al., *Preprint of Annual Meeting of Japan Soc. of Prec. Engg.* (1970), 109 (Japanese).
[21] Ikeda, M. et al., *Annual Report of FMSC in ETL* (1981), 21 (Japanese).
[22] Chesler, R. B. and Geusic, J. E., Ref. 17, vol. 1, (1972), 335.
[23] ibid., 349.
[24] Cheo, P. K., *Lasers*, Levine, A. K. and De Maria A. J., editors (Dekker, NY, 1971), 111.
Duley, W. W., *Co lasers* (Academic Press, NY, 1976), 59.
Metzbouer, E. A., editor, *Application of lasers in materials processing* (ASM, 1979), 15.
[25] Demaria, A. J., *Proceedings of the IEEE* **61** (1973), 731.
[26] Iwaki, K. et al., *Hitachi Hyoron* **64** (1982), 835 (Japanese).
[27] Takahashi, T. et al., Ref. 13, (1981), 111 (Japanese).
[28] Tabata, N. et al., *Proceedings of Int. Conf. of Appl. Lasers & Electro-Optics (ICALEO '84)* **44** (1984), 238.
[29] Jenkins, F. A. and White, H. E., *Fundamentals of Optics*, (4th edn, McGraw–Hill, 1976), 153.
[30] Ikeda, M., *Laser processing* (Keiei-system, Tokyo, 1980), 87 (Japanese).
[31] Jones, M. G. and Georgalas, G., *Laser News*, (July 1984), 11.
[32] Tsushima, K. et al., Ref. 28, 269.
[33] Sherman, G. H. and Frazier, G. F., *Optical Engg.* **17** (1978), 225.
Miyata, T., Ref. 19, **49** (1983), 1333 (Japanese).
Namba, H., et al., Ref. 28 **44** (Boston, 1984), 284.
Sakuragi, S. et al., ibid., 291.
[34] Sumiya, M. et al., ibid., 276.
[35] Ikeda, M. et al., Ref. 21 (1983), 7 (Japanese).
[36] Ikeda, M. et al., *Proceedings of ICALEO '83* **38** (1983), 16.
[37] Steen, W. M., *Optics and Laser Technology* **6** (1982), 149.
[38] Kobayashi, A., *Science of Machine* **30, 5** (1978), 583 (Japanese).
[39] Yasunaga, N. et al., *Bull. Japan Soc. of Prec. Engg.* **14, 3** (1980), 189.
[40] Allmen, M. von, *J. Appl. Phys.* **47, 12** (1976), 5460.
[41] Anthony, T. R., *J. Appl. Phys.* **51, 2** (1980), 1170.
[42] Kobayashi, A., *Laser machining* (Kaihatsu-shya, 1976), 90 (Japanese).
[43] Ohmine, M. et al., *Annual Report of National Project of FMSC* **V** (1979), 449 (Japanese).
[44] ———, ref. 28, (1984), 253.
[45] Hamada, K. et al., ref. 20, (1983), 309 (Japanese).
[46] Herbrich, H., *Rep. of Laser Institute of Japan (LIJ)* **4, 3** (1979), 6 (Japanese).
[47] Inagawa, T., *Handbook of laser application technology*, edited LIJ (Asakura, Tokyo, 1984), 187 (Japanese).

[48] Shimada, R., ref. 30, 12 (Japanese).
[49] Sekigichi, N., ibid. 142 (Japanese).
[50] Susuga, S. and Kutsuna, M., *Welding technique* **28, 1** (1980), 40 (Japanese).
[51] Kawasumi, H., *O plus E* **4** (1980), 29 (Japanese).
[52] Ohmine, M. et al., ref. 42, (1978), 349 (Japanese).
[53] Miyamoto, H., *Preprint of Symposium at Annual Meeting of Jap. Soc. of Prec. Engg.* (1981), 56 (Japanese).
[54] Ohmine, M. et al., ref. 42, (1981), 350 (Japanese).
[55] Sasaki, H., *OHM* **4** (1983), 44 (Japanese).
[56] Ikeda, M., Taikabutsu Overseas, **5** (1985), 27.
[57] Tsukamoto, K. et al., ref. 53, (1984), 19 (Japanese).
[58] Ikeda, M., ref. 49, **30** (1982), 28 (Japanese).
[59] Nagano, Y., ref. 46, 109 (Japanese).
[60] Shimoi, Y. et al., *Annals of the CIRP* **32, 1** (1983), 135.
[61] Miyauchi, T. et al., ibid. 141 and ref. 48.
[62] *Nikkei Mechanical*, (17 Jan 1983), 112 (Japanese).
[63] Bohman, C. F., *SME Technical Paper* **AD74-810**
[64] Gnanamuthu, D. G., *Optical Engg.* **19, 5** (1980), 783.
[65] Kawasumi, H. and Arai, T., ref. 19, **47, 12** (1981), 1470 (Japanese).
——, ref. 47, (1984), 150 (Japanese).
[66] Poete, J. M., *J. Phys. Colloq.* **41, C-4** (1980), 1.
——, *Semicond Silicon* (1981), 500.
[67] Draper, C. W., *Journal of Metals* (June 1982), 24.
[68] Hoshina, N. et al., ref. [42], (1980), 417 (Japanese).
[69] Kear, B. H., et al., *Metals technology* (April 1979), 121.
Pangborn, R. J. and Beaman, D. R., *J. Appl. Phys.* **51** (1980), 5992.
[70] Jau, B. M. et al., ref. 60, MR80–846
[71] Ikeda, M. et al., ref. 35, (1983), 134.
Yasunaga, N. et al., *Proceedings of 5th ICPE* (Tokyo, 1984), 478.
[72] Yasunaga, N., et al., *Proceeding of Intern. Conf. of Laser Advanced Materials Processing* (Osaka, 1987), 485.
[73] Copley, S. M. and Bass, M., *Proceedings of Conf. Appl. Lasers for Mater. Process* (1979), 121.
Copley, S. M., *Proceedings of ICALEO '82* (1982).
[74] Koyabu, K. and Watanabe, J., ref. 52, **1** (1985), 123 (Japanese).
[75] Haynes, R. W. et al., ref. 6, **37, 4** (1980), 344.
Chuang, T. J., *J. Chem. Phys.* **74, 2** (1981), 1453.
[76] Gutfeld, R. J. von and Hodgson, R. T., ibid. **40, 4** (1982), 352.
[77] Gutfeld, R. J. von et al., *Appl. Phys. Lett.* **35, 9** (1979), 946.
[78] Baghdadi, A. et al., *Applied Optics* **19, 6** (1980), 909.
[79] Ikeda, M., ref. 37, **30, 1 & 2** (1978), 9 & 238 (Japanese).
Gagliano, F. P. et al., ref. 25, **57** (1969), 114.
[80] *Optical Spectra* (July 1972), 18.
[81] Zehner, D. M. et al., ref. 6, **36, 1** (1980), 56.
[82] Takeda, G. et al., *Tool Engineer*, **25, 6** (1981), 38 (Japanese).
[83] Gagliano, E. P. et al., *Proc. IEEE* **57, 2** (1969), 2.

3
ELECTRON-BEAM PROCESSING

3.1 Introduction

3.1.1 Development of electron-beam processing

Sir William Crookes discovered in 1879 that a platinum anode is melted by bombardment with cathode rays. Thompson showed in 1897 that cathode rays are streams of electrons. Their application in industry was proposed by Pirani, who obtained a United States Patent for electron-beam melting in 1907. Ardenne indicated the possibility of using the corpuscular beam as a processing tool in 1938.

As for other types of electron-beam processing, Steigerwald pointed out its capability for machining in 1953. Stohr reported electron-beam welding in 1957, and Wyman also in 1958. A detailed chronology is given in Ref. [1].

Another field of electron-beam processing—electron reactive processing—was developed in the late nineteenth century; it was found that an electric discharge can initiate polymerization or depolymerization. Charlesby performed experiments on this phenomenon in the 1950s [2, 3]. This process has been applied in industry, using effective electron beam sources.

3.1.2 Effects of electron bombardment on materials

When energetic electrons are projected onto solid materials, some are reflected from the surface layer by elastic or inelastic collisions. However, most of the electrons penetrate the solid material, losing their kinetic energy. During this process, complex phenomena due to interaction with the atoms of the solid occur, as shown schematically in Fig. 3.1 [4]. The solid surface layer can be examined using secondary or reflected electrons, X-rays and fluorescence, as in scanning electron microanalysis, X-ray microanalysis, fluorescence microanalysis, etc.

This chapter is mainly concerned with thermal processing by electron-beam heating and chemical reaction of the molecules or atoms of the workpiece. Practical applications are as follows:

(i) thermal processing: machining, welding, annealing, heat treatment, etc.
(ii) reactive processing: electron-beam lithography, electron-beam polymerization and depolymerization, etc.

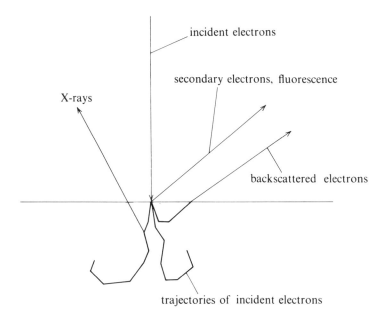

FIG. 3.1. Schematic diagram of bombardment effects of electrons [4]

(1) Thermal processing

Electrons projected onto a solid material penetrate the surface layers to some depth. Whiddington showed experimentally that the electron penetration range R_p (m) is determined by the electron acceleration voltage V (V) and the mass density ρ (kg/m^3) of the metal [5, 6]. The relation is expressed by $R_p = 2.2 \times 10^{-11} V^2/\rho$.

The energy loss along the electron penetration trajectory has been investigated theoretically [7–10], experimentally [11–13] and by computer simulations based on the Monte Carlo technique [14, 15]. Relations between energy loss in unit penetration mass-thickness and non-dimensional depth are shown in Fig. 3.2 [10], where E_A, E_0, ρ, R_p, and z are the absorbed energy, incident energy of electrons, mass density of the material, maximum penetration range, and depth from the material surface, respectively. The peak value of energy loss occurs at a certain depth in the target material.

Figure 3.3 shows trajectories of incident electrons obtained by the Monte Carlo technique [15]. Projected electrons scatter in all directions inside the material. A finely focused electron beam supplied to the workpiece surface is scattered fairly widely inside the workpiece, so the beam cannot process a very small area, even though the diameter of the beam is very small.

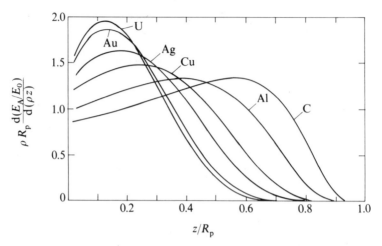

FIG. 3.2. Normalized fraction of energy dissipated in unit mass-thickness as a function of reduced depth z/R_p [10]

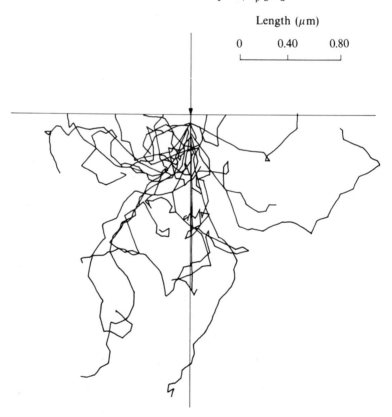

FIG. 3.3. Trajectories of 30 keV electrons in copper by Monte Carlo technique based on multiple scattering [15]

(2) Reactive processing

During the scattering of incident electrons, secondary electrons are generated, and the ionization and excitation of constituent molecules of the material occur as follows [16–18]:

$$M \begin{cases} \rightsquigarrow M^+ + e \\ \rightsquigarrow M^* \end{cases}$$

where $\rightsquigarrow$, M, M$^+$, M*, and e represent the direction of the energy absorption, molecule, ion, excited molecule, and electron, respectively. The excited ion has a short lifetime and becomes an excited molecule:

$$M^+ + e \rightarrow M^*$$

Some excited molecules lose their energy by collision with other molecules; some are changed into radicals:

$$M^* \rightarrow R_1\cdot + R_2\cdot$$

When molecules have a tendency to combine with electrons, negative ions are formed:

$$M + e \rightarrow M^-$$

The generated M*, M$^+$, M$^-$, $R_1\cdot$, $R_2\cdot$, and e are called active species and induce chemical reactions. Electron-beam lithography is based on this process, and so is electron reactive processing.

3.1.3 *Concept of and equipment for electron-beam processing*

Electron-beam processing is defined as a technique which changes the shape or properties of a material or workpiece by the use of electron beams. The useful features (i)–(iv) and disadvantages (v)–(vii) of electron-beam processing are as follows:

(i) possibility of finely focused electron beams
(ii) feasibility of generating high-power-density electron beams
(iii) ability to deflect electron beams rapidly and highly accurately
(iv) possibility of varying electron energy with acceleration voltage and hence of controlling electron penetration range
(v) necessity for high vacuum to generate electrons and for work chamber (except for non-vacuum welding and electron reactive processing)
(vi) generation of harmful X-rays
(vii) some difficulty in processing electrical insulators.

The drawbacks (v) and (vi) may be converted into useful features in some cases, because (v) means that the electron-beam processing is performed in a clean environment and (vi) that in-process monitoring is possible using the generated X-rays as a sensor. Of course, these two drawbacks are unimportant in automated facilities.

The basic equipment of electron-beam processing is the same as that of electron microscopy. In practice it is relatively simple, as already shown in Fig. 1.10. The equipment consists of electron gun, focusing coil, deflecting elements, work chamber, vacuum system, and power supplies. The electron gun consists of a cathode which generates electrons, a control electrode, and an anode; its working principle is the same as that of the triode vacuum tube. The cathode is set at a potential up to about 150 kV, and the anode is set at earth potential. The electrons are accelerated through this potential field, and the accelerated beam is focused by the magnetic coils and deflected by the magnetic or electrostatic fields of the deflecting elements. The potential difference between the cathode and the control electrode is of the order of 10^1 to 10^2 V; the beam intensity is easily controlled by changing this potential difference.

The machine shown in Fig. 1.10 is a simple version; when a fine, symmetrical beam is required, aperture, stigmator or multifocusing coils are included. Furthermore, there are many types of cathode; a general type is a heated hairpin filament. When a high-current beam is required, strip-type, block-type or plasma hollow cathodes are used. When a low current and fine beam are required, an LaB_6 tip-type cathode is used; this is sometimes used in static field emission.

3.2 Electron-beam machining

This type of process is carried out by stock removal from a localized portion of the workpiece by melting and/or vaporization due to localized heating by a high-power-density electron beam. This process was initially expected to be applied successfully in fully automated factories, because the electron beam can be easily and accurately controlled automatically and electron-beam power is generated at very considerably high efficiency. However, nowadys its use in industry is limited.

This section describes mainly drilling, which is the main type of stock-removal electron-beam process, but processes such as trimming and cutting will also be described.

3.2.1 Drilling

(1) Mechanism

Stock removal by a high-power-density electron beam is carried out by melting and vaporization of the material. During the drilling of a deep hole, the ejection of melted material from the surface of the workpiece can be seen with the naked eye. As mentioned above, high-energy electrons penetrate the surface layer of the workpiece; thus the temperature rises throughout the penetration depth [19–21]. A surface layer of some thickness is melted simultaneously as a result of this penetration and thus bubbles are generated [22, 23]. Incident electrons then pass through the bubble and heat the molten zone around it. The rise in vapour pressure inside the bubble causes it to burst, and the surrounding melted material is ejected, as shown in Fig. 1.13. Repetition of this process is the basis of drilling.

The quantities removed in the molten and vapour states have been calculated as shown in Table 3.1 [20]. Almost all the metal removed is in the molten state.

TABLE 3.1

Quantities removed in vapour and molten states in the initial stage of electron-beam drilling (150 kV, 40 mA, 0.5 mm diameter) [20]

	Aluminium	Copper
Vaporized mass (g)	5×10^{-8}	1×10^{-7}
Melted mass (g)	3×10^{-5}	5×10^{-5}

In drilling unfired ceramics containing organic binder, fairly large quantities of gases are generated [24] and the expansion of these gases causes the ejection of solid matter [25].

Examples of drilled holes are shown in Fig. 3.4.

(2) Focusing conditions

The electron beam is focused on the surface of the workpiece by magnetic lenses; hence the focal length appropriate for drilling can be adjusted by changing the lens current according to the dimensions of the machine. The power density, drilled depth, and diameter vary with the lens current [26–28]. There is a tendency for the deepest hole to be drilled when the diameter of the hole is smallest. In practice, the appropriate focusing condition can be determined experimentally by measuring the inlet diameter of the hole.

The above discussion considers mainly drilling of thick plate. Different focusing conditions must be chosen for piercing a thin plate or foil. The appropriate focusing condition is defined as that under which the electron beam can drill a hole through the plate using minimum energy, i.e. the shortest pulse duration. In this case, the appropriate focusing lens current depends on the thickness of the plate or foil [29].

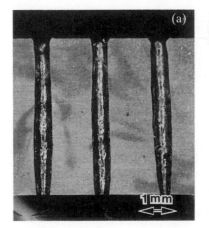 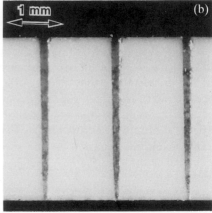

FIG. 3.4. Holes drilled by 120 kV single-pulse electron beams (Courtesy Hitachi, Ltd). (a) In stainless steel 304 (5 mm thick). (b) In Al_2O_3 ceramic (3.2 mm thick)

(3) Pulse conditions

Drilling may be carried out by single or multiple pulses. This section deals with the effects of pulse duration and number on the drilling characteristics.

(a) Pulse duration A hole is produced by a sequence of stock removals. In the initial stage of the drilling, the depth increases with the pulse duration; however, a saturation level is reached at a certain pulse duration. In a hole of saturation depth, a fairly thick resolidified melted zone can be seen around the wall. This means that the molten material is not ejected from the hole. The process is shown schematically in Fig. 3.5 [20]. With a short pulse for a shallow hole depth, the molten material is ejected from the hole completely (Fig. 3.5(a)). As the hole becomes deeper, the molten material formed at the

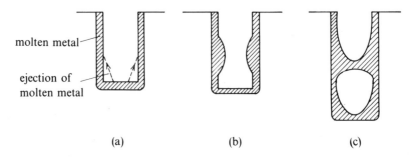

FIG. 3.5. Drilling process in metals [20]

end of the hole can no longer be ejected and becomes attached to the side wall (Fig. 3.5(b)). The attached material contracts owing to the surface tension and fills part of the hole (Fig. 3.5(c)). That is, for deeper holes, the electron beam no longer has the ability to remove stock under these conditions. At this stage of drilling, the quantity of melted material remaining increases, and welding occurs. To drill a hole, it is necessary to use a pulsed beam of duration less than this threshold.

The dependence of hole depth on pulse duration is illustrated in Fig. 3.6 [20]. Under these conditions (150 kV, 40 mA, about 3×10^{10} W/m^2), the threshold duration is about 1 ms.

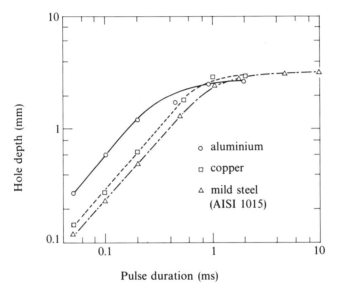

FIG. 3.6. Dependence of hole depth on pulse duration obtained by single pulse [20]

The drilled depth is proportional to the pulse duration for short durations. If stock removal in the molten state is taken into account, the drilling speed dD/dt, where D is the drilled depth and t the time, for the one-dimensional case in a semi-infinite body, is given [20, 30] by

$$\frac{dD}{dt} = \frac{Q}{\pi a^2 \rho (L_m + cT_b)} \qquad (3.1)$$

where Q is the input electron beam power (W), a the radius of the circular-section electron beam (m), ρ the mass density (kg/m^3), L_m the latent heat of fusion (J/kg), c the specific heat (J/(kg K)), and T_b the temperature of the ejected molten metal. This formula will be discussed in Appendix A

TABLE 3.2
Initial drilling speed (m/s) under the same conditions as in Table 3.1

	Aluminium	Copper	Mild steel (AISI 1015)
Experimental	5.4	2.6	2.4
Calculated from eqn. 3.1	4.4	2.5	2.1

(eqn. A.21). The calculated results deduced from eqn. 3.1 and the experimental results obtained from Fig. 3.6 are listed in Table 3.2.

(b) Number of pulses When the required depth cannot be obtained by a single pulse, multiple pulses are necessary. The depth, however, does not increase linearly with the number of pulses and has a tendency to reach a

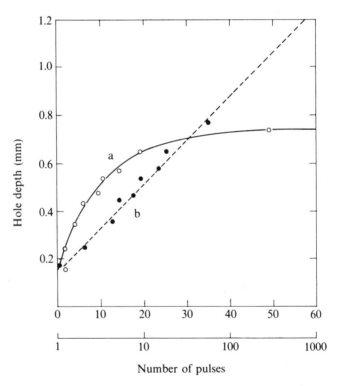

FIG. 3.7. Dependence of hole depth on number of pulses. **(a)** Linear scale. **(b)** Logarithmic scale [32]

saturation level [31, 32], as shown in Fig. 3.7 [32]. Drilling was carried out by an electron beam at 130 kV, 6 mA, and pulse duration 10 μs.

The hole diameter, on the other hand, depends not on the number of pulses but on the pulse duration [27]. The drilling conditions must therefore be determined from the required hole shape. When the hole diameter is limited, the pulse duration must be selected to obtain the desired diameter. When a deep hole whose diameter has no limitation is required, long pulses may be used.

The total energy consumed in constant-power drilling is determined by the product of pulse duration and number of pulses. Below a certain energy value, the maximum depth is obtained by the longest pulse, i.e. a single pulse [27].

(4) Properties of material

A non-dimensional formula for drilled depth is deduced from eqn. 3.1 as follows:

$$\frac{\lambda(T_m + L_m/c)D}{Q} = \frac{\kappa\tau}{\pi a^2} \qquad (3.2)$$

where the melting point T_m is used instead of T_b, and κ is the thermal

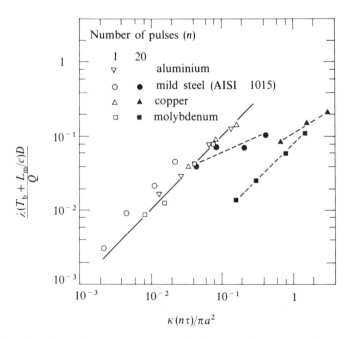

Fig. 3.8. Non-dimensional relationships for hole depth obtained by single and multiple pulses [27]

diffusivity, given by $\lambda/\rho c$. When the number of pulses is n, τ is replaced by $n\tau$. For $n=1$ and $n=20$, the relationship of eqn. 3.2 is shown in Fig. 3.8 [27]. For $n=1$, the non-dimensional formula represents the drilling characteristics for various metals by a single line, but this is not true for multiple pulse drilling.

(5) Drilling range

Drilled depth, diameter, and rate of hole production (holes per second) are of practical importance [33–36]. Figure 3.9 shows the relations for steel and nickel alloys [34]. The available value of depth/diameter is about 10, and 500 holes of 0.1 mm diameter can be drilled in a 1 mm plate within 1 s. These drilling characteristics depend largely on the machine capacity.

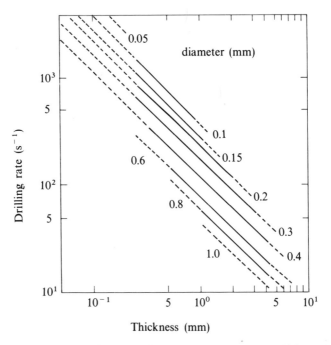

FIG. 3.9. Drilling rate as a function of hole diameter and plate thickness (steel and nickel alloys) [34]

(6) Control of hole diameter at rear surface of a plate

When a small stock quantity is removed by each pulse, it is possible to control the hole diameter at the rear surface of the pierced plate by changing the pulse conditions. Examples are shown in Fig. 3.10 [37], where the total energy consumed in drilling is shown in parentheses. Pulse conditions can be

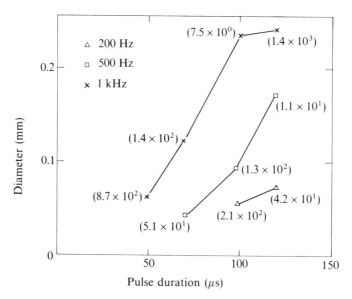

FIG. 3.10. Hole diameter at rear surface of copper plate 0.5 mm thick obtained by an electron beam of 100 kV and 30 mA (values in parentheses are energy (J) consumed in drilling) [37]

controlled easily, so this control method appears to be important in electron-beam drilling. The necessary number of pulses, however, increases rapidly with decreasing pulse duration [31, 32].

3.2.2 Cutting of plates

Electron-beam cutting of plate several millimetres thick was carried out in the early stages of the development of electron-beam machining [38, 39]. For example, an 8 mm plate of 18–8 nickel–chromium steel was cut by an electron beam of 1.6 kW at a speed of 120 mm/min [38]. Electron-beam cutting in non-vacuum also has been developed; stainless steel 6 mm thick can be cut by an electron beam (110 kV, 11 mA) at a speed of 100 mm/min using auxiliary helium gas [40]. This kind of cutting process, however, has not been successfully used in industry.

3.2.3 Machining of thin film and foil

Electron-beam machining can be done without mechanical forces except for the recoil force due to stock removal [41]. When thin foils are machined, the machining rate is very high (see Fig. 3.9). By deflection of the beam, a complex pattern can be machined at high speed and with high accuracy.

For slits machined in various materials, it is reported that a narrower slit width can be obtained with a lower energy [42]. When foils thinner than the electron penetration range R_p are machined, the foil thickness has a strong effect on the machining characteristics, because some of the electron-beam power passes through the foil.

3.2.4 Other machining processes

Electron-beam machining can be widely applied in trimming, scribing, and marking in the electronics and semiconductor industries [43, 44], because it can be carried out more easily and with higher accuracy and higher speed than by a laser beam. However, electron-beam techniques need more elaborate systems and have a higher initial cost than laser-beam methods and consequently have not so far been fully applied in industry.

Although mainly the machining of metals is treated here, ceramics can be easily machined, as shown in Refs. [25, 26, 45] (see Fig. 3.4).

3.3 Electron-beam welding

The most important feature of electron-beam welding is that narrow and deep welding is possible. A depth–width ratio of about 10 or more is easily obtained. Applications include a wide range of joining techniques. In this section, the fundamentals of electron-beam welding are discussed.

3.3.1 Mechanism of penetration welding

Deep-penetration welding of steel to a depth of 150 to 200 mm has recently been carried out. The penetration range R_p of electrons into solid or liquid metals is of the order of 10^1 μm at 100 keV. The weld zone due to direct heat conduction should be approximately hemispherical; therefore, the mechanism of deep welding cannot be explained in these terms.

For stationary electron-beam welding, where there is no relative movement between electron beam and workpiece, a mechanism involving formation of a long, narrow cavity has been proposed [46]. Through this long cavity filled with vapour and molten metal, electrons can penetrate deeply into the workpiece material and reach the bottom of the fusion zone. However, this kind of long cavity does not seem to exist stably in the fusion zone. Therefore, an alternative model of a chain of bubbles has been proposed, as shown in Fig. 1.14 [47].

In actual deep welding, such a cavity moves in the molten material and forms a fusion zone along the path of the moving electron beam, as shown in

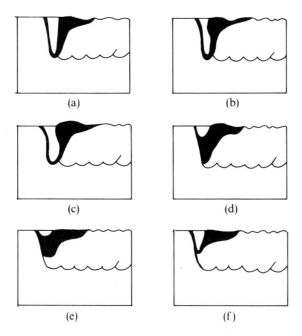

FIG. 3.11. Cavity oscillation sequence for 1100 aluminium [48]

Fig. 3.11(a). During the movement, the cavity shape is not stable but fluctuates with a certain frequency (Fig. 3.11(b)–(f)) [48]. The forces acting on the fusion zone [49–53] and the heat balance and molten metal flow [54] have been evaluated. Molten metal flow has been observed by pulsed X-ray photography [48], pinhole camera photography using generated X-rays [55], measurement of generated X-rays by film and scintillation counter [56, 57], and dynamic fluoroscopy using projected X-rays [58]. The molten metal existing behind the cavity moves upwards with a speed of 9 m/min [58], and analysis of this movement indicates the existence of a vortex [59].

A dynamic analysis has been made of cavity oscillation, assuming a frictionless fluid, and the oscillation frequency has been predicted [48]. In full penetration of thin plate, or in the forward half of the molten pool in deep penetration, the molten metal flow in the direction of penetration may be considered negligible in the two-dimensional liquid flow. For such a case, liquid flow has been analysed on the basis of the surface tension force produced by the temperature gradient around the cavity [60]. However, the molten layer thickness has not yet been confirmed. The thickness distribution of the molten layer around the cavity has been calculated, as shown in Fig. 3.12 [60]. The quantity S is the reduced (dimensionless) surface tension defined by $C_\sigma(T_m - T_0)/\mu U$, where C_σ is the temperature coefficient of surface

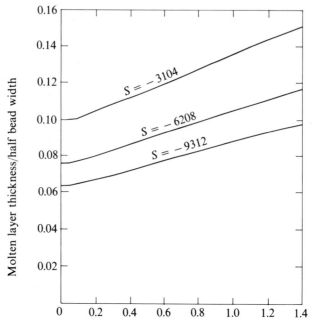

FIG. 3.12. Non-dimensional molten layer thickness in aluminium for $Re = 36.4$ [60]

tension (N/m K), T_m the melting point (K), T_0 a reference temperature (K), μ the viscosity (Pa s), and U the welding speed (m/s); ξ is the non-dimensional length defined by l/b, where l is the length from the leading point of the cavity to the calculation point along the molten layer boundary, and b is the minor axis of the cavity, which is assumed to be elliptical.

For the welding of aluminium at $U = 18$ mm/s and Reynolds number $Re = Ub/v = 36.4$, where v is the kinematic viscosity (m²/s), the value of S is -6208. For a typical cavity of minor axis $b = 1.5$ mm, the molten layer thickness at $S = -6208$ varies from 0.1 at the leading point, $l = 0$, to 0.18 mm at the side wall.

3.3.2 Penetration-welding depth

(1) Factors affecting penetration depth

Penetration depth depends on various operating variables of the electron beam, such as acceleration voltage, beam current, beam diameter, welding speed, beam focus position, degree of vacuum, material properties, and welding position [61–65]. Figures 3.13 and 3.14 show its dependence on beam

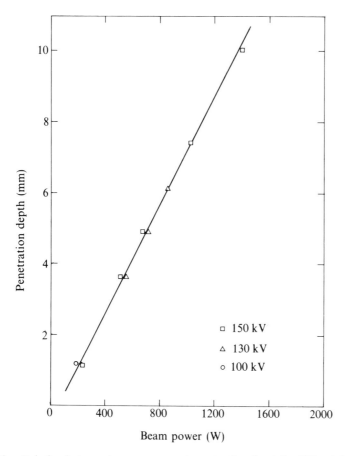

FIG. 3.13. Relation between beam power and penetration depth for 302 stainless steel (welding speed = 686 mm/min) [61]

power and thermal conductivity of the material under a vacuum of 7 mPa [61]. Generally a shallow penetration is obtained in materials of high melting point, high latent heat of fusion, high thermal conductivity, and low vapour pressure, and at high welding speed [62].

Penetration depth and bead width vary with the focus of the beam, usually defined by the parameter $a_b = D_0/D_f$ [67], where D_0 is the length from the focusing lens to the workpiece surface (working distance), and D_f the length from focusing lens to the focus position or cross-point of the beam (focal length). The optimum value of a_b at which the deepest penetration can be obtained depends also on the welding conditions and should be determined from them.

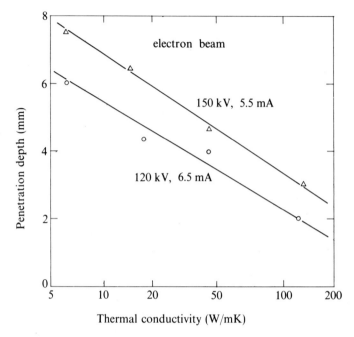

FIG. 3.14. Relation between thermal conductivity and penetration depth for various metals (welding speed = 686 mm/min) [61]

The degree of vacuum is strongly related to the convergence of the beam; electrons are scattered by residual gas molecules, and a finely focused beam cannot be obtained. The dependence of penetration depth on the degree of vacuum is shown in Fig. 3.15 [63]; above 1.3 Pa, the depth decreases rapidly with increasing pressure.

In non-vacuum welding, considerable scattering of electrons occurs in the surrounding gas [68]; to reduce this, the distance from the exit of the electron beam to the workpiece should be as short as possible. Of course, scattering depends on the kind of gas; helium can reduce the scattering and increase the penetration depth by a factor of two in comparison with air.

The welding position of the workpiece largely determines the stability of the cavity. For shallow penetration, less than 35 mm, there is no difference in penetration depth between five positions [69]: flat; vertically down with workpiece motion and gravity in opposite directions; vertically up with workpiece motion and gravity in the same direction; overhead; and horizontal–vertical with both the electron beam and workpiece motion perpendicular to gravity. However, in the overhead position, molten metal flows out and droplets are formed. The value of a_b also affects the penetration

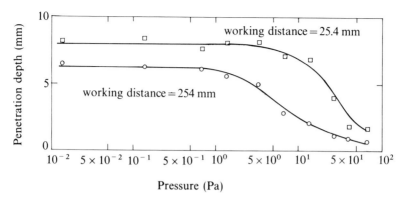

FIG. 3.15. Effect of pressure on penetration depth (100 kV, 9 mA electron beam; 304 stainless steel) [63]

depth. For a large value of a_b, deeper penetration is obtained in the horizontal–vertical position than in the flat position [70].

Above a penetration depth of about 100 mm, a position effect occurs. The cavity cannot exist stably because of the liquid gravity head in the molten pool. Penetration deeper than 125 mm cannot be obtained in the flat position; however, penetration deeper than 200 mm may be obtained in steel by 75 kW (150 kV, 500 mA) horizontal welding [71].

(2) Formulae for penetration depth of electron-beam welding

Exact formulae including all the parameters have not been obtained. Formulae for penetration depth H (m) obtained from dimensional analysis [72–74], heat conduction theory [75–78], and experimental analysis [79] are as follows:

$$H \propto \frac{Q}{\lambda T_m} \sqrt{\left(\frac{\kappa}{Ud}\right)} \qquad (3.3)\,[72]$$

$$H \propto \frac{Q}{\lambda T_m} \left(\frac{\kappa}{Ud}\right)^{0.83} \qquad (3.4)\,[79]$$

$$H = \frac{2Q}{Ud\rho(cT_m + L_m)} \frac{1}{1 + 5\lambda\left(\frac{1}{d} + \frac{U}{2\kappa}\right) \frac{T_m}{\rho(cT_m + L_m)} \frac{1}{U}} \qquad (3.5)\,[75]$$

where Q is the total input power (W), λ the thermal conductivity (W/m K), ρ the mass density (kg/m^3), c the specific heat (J/kg K), L_m the latent heat of fusion (J/kg), T_m the melting point (K), κ the thermal diffusivity (m^2/s), U the welding speed (m/s), and d the electron-beam diameter (m). The beam diameter d cannot be determined precisely, because the beam power density

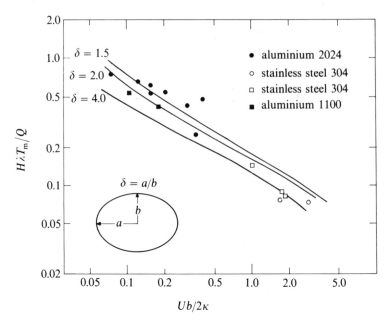

FIG. 3.16. Non-dimensional relations for penetration depth in partial penetration welding [80]

distribution varies with position along the beam path, generally obeying the Gaussian distribution. Furthermore, electrons are scattered by vaporized gas inside the cavity in the weld zone. It is therefore difficult to make definite comparisons between measurements by different investigators.

A method of overcoming this difficulty is to consider the weld width instead of the beam diameter. All the energy deposited in a cavity in the weld zone will be lost by conduction to the surrounding solid. On the basis of this fact, and using the heat transfer from an elliptical cylinder moving through an infinitely wide plate and whose boundary temperature is kept constant, non-dimensional relations for weld width $2b$ and penetration depth H are obtained as shown in Fig. 3.16 for various values of δ, where δ is the ratio of the major and minor radii of the elliptical cylinder of the melted region [80]. The value of δ is expected to depend on the material and welding conditions. However, a mean curve corresponding to $\delta = 2.0$ may be adequate for general use.

3.3.3 Properties of the weld zone

The weld zone produced by electron-beam welding has properties different from those produced by other welding techniques. Oxidation and gas absorption in the weld zone do not occur, because of the high vacuum. The weld zone, distortion of the workpiece, and heat-affected zone are comparatively small because of the low input energy and high welding speed.

There is a greater possibility of welding dissimilar metals of largely different thermal properties, because the welding mechanism is based not on heat conduction but on direct heating by electrons [81].

(1) Mechanical properties

The mechanical properties of the weld zone depend on the kind of material [81, 82]. For example, carbon steel is hardened; on the other hand, cold-rolled 300-series stainless steel is softened [83–86]. The hardness of electron-beam-welded structural steel has been predicted by applying the weld-zone cooling rate to the CCT (continuous cooling transformation) curves for welding, as shown in Fig. 3.17 [87]. A formula for predicting the hardness in the weld zone has also been derived [88]. The mechanical strength is, in general, the same as that of the base material; however, the tensile strengths of cold-rolled metals and some titanium alloys are reduced [89].

The welding characteristics of various metals are listed in Table 3.3 [90].

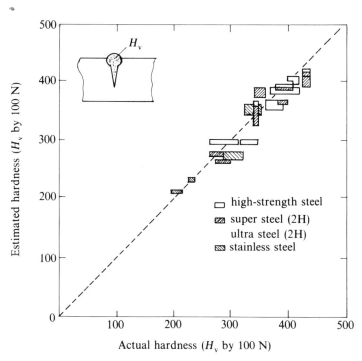

FIG. 3.17. Relation between actual and estimated hardness of weld for structural steel [87]

TABLE 3.3
Quality rating of various metal welds [90]

	Zr2Sn	V10Ti	V	Ti	Ni15Cr7Fe	Ni	Nb	Fe18Cr8Ni	Cu20Ni	Cu	Al	Ag
Ag	O	O	●	●	●	O	O	●	●	O	●	O
Al	O	O	O	O	⊗	●	O	⊗	⊗	⊗	O	
Cu	O	⊘	●	O	●	⊘	O	●	O	O		
Cu20Ni	O	⊗	⊗	O	●	O	O	●	O			
Fe18Cr8Ni	O	O	⊘	O	●	●	O	O				
Nb	⊗	●	●	●	O	O	O					
Ni	O	O	O	⊗	●	O						
Ni15Cr7Fe	O	⊗	⊗	O	O							
Ti	●	O	⊗	O								
V	O	O	O									
V10Ti	O	O										
Zr2Sn	O											

● Good
⊘ Fair
⊗ Poor
O Not investigated

(2) Defects

Several defects may arise in the electron-beam weld zone, such as porosity, spiking, cracking, and cold shuts, as shown in Fig. 3.18.

Porosity is caused by gaseous components or volatile elements in the material [84]; this may be prevented by oscillation of the electron beam along the welding direction [91, 92].

Spiking is a result of irregular penetration of the electron beam [48, 57] and is considered a defect when it becomes too large. The spikes usually exhibit a

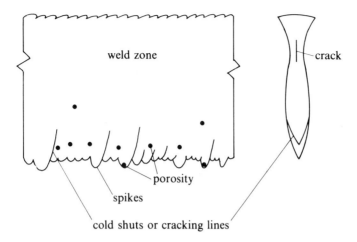

FIG. 3.18. Defects in electron-beam welding

periodic pattern whose frequency has been theoretically analysed (see Fig. 3.11)[48]. Spikes are likely to occur in welding at high power density [93].

Cracking occurs as a result of non-uniform solidification of the molten metal. This may be prevented by selecting alloys of the correct component ratios, setting the correct focusing position, and reducing the welding speed.

Cold shuts at the bottom of the weld region are due to insufficient bonding of melted material to the base material. This may be prevented by reducing the cooling rate and oscillating the beam. The occurrence of this defect is also related to the focusing position and the oscillation frequency of the beam, as shown in Fig. 3.19 [91].

3.3.4 Equipment

Equipment for electron-beam welding may be classified as follows [94–96]:
 (i) Classification based on processing environment:
    (a) high-vacuum type ($\gtrsim 10^{-2}$ Pa)
    (b) partial-vacuum type ($\simeq 10^0$ Pa)
    (c) non-vacuum type (atmospheric).
(ii) Classification based on acceleration voltage:
    (a) high-voltage type (100–150 kV)
    (b) low-voltage type (30–60 kV).

The partial-vacuum type is used for high productivity because of the time saved in preparing a suitable welding environment. The non-vacuum type is properly used for mass production and sequential welding at a high rate.

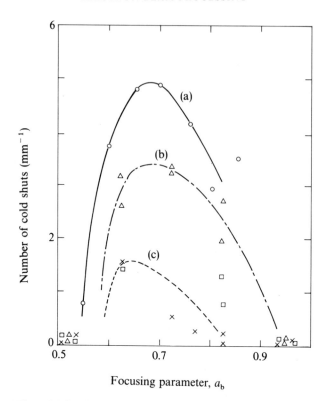

FIG. 3.19. Effect of deflection of electron beam on cold shut generation. (a) Stationary beam. (b) Oscillating deflected beam, $f = 500$ Hz, amplitude of oscillation = 0.6–1.6 mm. (c) Oscillating deflected beam, $f = 5000$ Hz, amplitude of oscillation = 0.8–2.4 mm [91]

These types of welding cannot be expected to give the same properties as those produced in the vacuum environment.

The high-voltage type is used for precision welding using a fine beam of low current. The low-voltage type in general use is furnished with an electron gun which can move along the welding path and is operated at a short work distance to avoid divergence of the beam.

Figure 3.20 shows a machine of partial-vacuum type which is equipped with multiple chambers; the evacuation time for preparing a working vacuum is not necessary because of the sequential feeding of the workpiece into the operating position.

The equipment above is of the type in which the workpiece to be welded must be transferred to the machine. Large parts and structures are therefore difficult to weld. A local-vacuum chamber type of machine has been

Fig. 3.20. Partial-vacuum electron-beam welder (Courtesy NEC)

developed [97–99]; a vacuum is produced only at the position to be welded, using special devices such as sealing tapes.

3.3.5 Applications

In the early stages of its development, electron-beam welding was applied to special fields, such as the atomic and space industries, neglecting economic considerations. At present, however, its application has been extended to the automobile and machine part industries, as follows:

(i) welding of thin plate and machine parts in automobiles, machines in general, airplanes, electronics, tools, etc.
(ii) welding of thick plates and large structures in heavy industries
(iii) welding of special materials in the atomic, space, and chemical industries.

Examples of welded parts are shown in Fig. 3.21.

Ceramics with high electrical resistivity and thermal fragility are not easily welded, but the process is not impossible [100]; there have been a few applications and investigations.

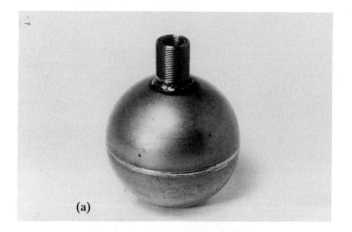

Fig. 3.21. Examples of electron-beam welding (Courtesy NEC). (a) High-vacuum welding of titanium hemispheres and tube. (b) Partial-vacuum welding of parts of propeller shaft

3.4 Electron-beam lithography

Lithography is a very important technique in manufacturing integrated circuits [101, 102]. This processing technique has been widely used in the fabrication of microcircuits such as LSI and VLSI. The function of the electron-beam lithographic process in LSI technology is to transfer very fine patterns onto the surface of semiconductor wafer, coated with thin polymer film (electron-beam-sensitive resist). The patterns transferred are developed by subsequent soaking of the wafer in a solvent.

Compared to photon lithography, electron-beam lithography has the advantage of being able to produce extremely fine patterns with high accuracy, because the equivalent wavelength of the electron beam is considerably shorter than that of a light beam. As a result, higher resolution can be obtained. Further developments in high accuracy (overlay, linewidth, positioning) and higher productivity (wafers per hour) are expected.

3.4.1 The process

Electron-beam lithography is used in the following fields:

(i) production of master mask (magnification = 1) for other lithographic processes
(ii) production of reticle mask (magnification = 10) for other step-and-repeat lithographic processes
(iii) direct writing of patterns on the wafer.

These processes are outlined in Fig. 3.22. The masks (a) and (b) can be produced by a pattern generator using visible or ultraviolet light, but with the increase in degree of integration or density of LSI, direct writing using the electron beam has been adopted for higher accuracy and productivity.

The fundamental processes are shown in Fig. 3.23. There are two types of electron resist, i.e. positive and negative, according to the generic type of chemical reaction. Positive resists are polymers whose molecular weight is reduced in the irradiated area and can be dissolved in a solvent that cannot

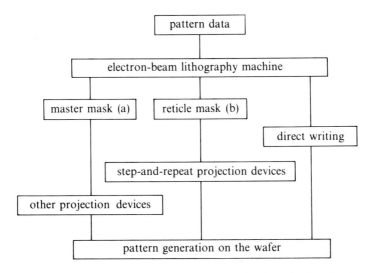

FIG. 3.22. Electron-beam lithography for fabrication of microcircuits

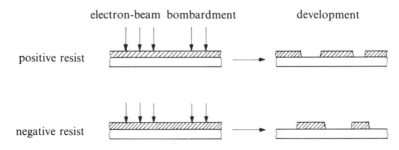

FIG. 3.23. Types of resist for electron-beam lithography

attack high molecular weight materials; negative resists are polymers whose molecular weight is increased by cross-linking and cannot then be dissolved in the solvent.

As a positive resist, poly(methyl methacrylate) (PMMA) is the most widely used for experimental and high-resolution lithography. It has a resolution of submicrometre order, but exhibits a rather low sensitivity of $\simeq 10^{-5}$ C/cm^2 [103]; it is not of great practical use. As a negative resist, COP (copolymer of glycidyl methacrylate and ethyl acrylate) exhibits a high sensitivity of $\simeq 10^{-7}$ C/cm^2 [103] but slightly lower resolution of about 1 μm, so, it is not adequate for VLSI fabrication.

Many electron-beam resists have been developed, and new resists are under research and development. Examples are listed in Table 3.4 [104].

3.4.2 Equipment

Electron-beam lithography is used for much finer patternmaking than are other thermal electron-beam processes. The construction of the machine is almost the same as shown in Fig. 1.10, but the electron-beam lithography machine must be equipped with components of much higher accuracy and stability than an ordinary one.

(1) Components

(a) Electron sources In the early stages, the electron source was a heated tungsten filament, because the machine was developed from the SEM (scanning electron microscope). Even though a tungsten filament operates stably in a relatively low vacuum, it is not suitable for IC production; it takes too long to write fine patterns because of the low electron emission. To extend the life of tungsten filament cathodes, thoriated tungsten filaments are used because they can emit the same current at a lower temperature because of their low work function [105].

Recently, cathodes with brighter emission than the tungsten cathode, such

TABLE 3.4
Sensitivity of resists for electron beam lithography [104]

Resist	Tone	Sensitivity[†] ($\mu C/cm^2$)
FBM	Positive	0.4
FPM	Positive	5
MPR	Positive	3
EBR–1	Positive	3.8
EBR–9	Positive	0.8
PMMA	Positive	16
FMR	Positive	0.2
PMMA–AN	Positive	5
PMIPK	Positive	2
PBS	Positive	0.7
OEBR–10	Negative	0.5
SEL–N	Negative	0.8
CMS	Negative	0.4
COP	Negative	0.5

[†]For 10 keV electrons

as LaB_6 (lanthanum hexaboride) and TFE (thermally assisted field emission) cathodes for electrostatic field emission and thermal emission, have been developed [105]. Though these sources are more difficult to handle and to fabricate than the tungsten cathode, they are useful for obtaining higher productivity and finer patterns.

(b) Electron optical and deflection systems Patterns in electron-beam lithography are produced by scanning the image of the cathode at the cross-over point or the aperture image on the wafer. Therefore, the machine needs electron imaging and demagnifying systems, which in general are magnetic lenses, and a magnetic or electrostatic deflection system.

Magnetic deflection causes quite a small aberration, but the rate of deflection is reduced by eddy currents and hysteresis. Electrostatic deflection is rapid but produces a large aberration. Various methods have been developed to overcome the difficulty, such as a separate movement method (where larger movement is effected magnetically and smaller movement electrostatically) [106], an electrostatic octopole deflector which is capable of deflection over the full 5 mm pass [107], and a 20-pole deflector which provides large field (10 mm) deflection [108, 109].

(2) Machine

There are several types of machine, classed according to the various combinations of shape of electron beam, scanning method, and method of movement of wafer.

(a) Shape of electron beam There are two types of beam in general: Gaussian round beam and shaped beam. The former is obtained by projecting the image at the cross-over point of the electron gun, which is used for SEM. The shaped beam is obtained by projecting the aperture image of rectangular and variable shape. This shaped beam emits a higher current than the Gaussian beam and can reduce the exposure time. The electron optical concept for a variable-shaped beam is shown in Fig. 3.24 [110].

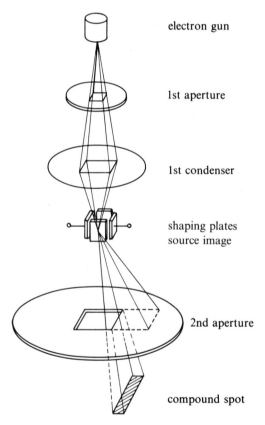

FIG. 3.24. Electron optical concept for variable-shaped beam [110]

(b) Scanning method Two methods are used in general. One is raster scanning, in which the electron beam scans the whole area of the field and irradiates the required portion, but time is lost in scanning unnecessary area. The frequency and amplitude of deflection are fixed and thus aberration is small. The other method is vector scanning, in which the electron beam scans only the exposure area. These are shown schematically in Fig. 3.25. A hybrid scanning mechanism has also been developed.

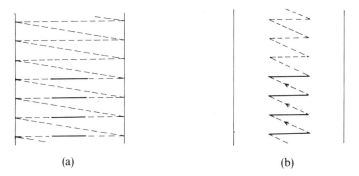

FIG. 3.25. Schematic diagrams of scanning methods. (**a**) Raster scanning. (**b**) Vector scanning

(c) Working-stage movement The stage of the wafer must be moved because the scanning range is not large enough to cover the whole area of the wafer. There are two movement methods: step-and-repeat; and continuous. The movement must be accurately measured and checked by laser interferometer, or by location registration marks on the wafer within each pattern area, using reflected electrons.

Examples of machines with different combinations of these types of mechanism are listed in Table 3.5 [111]. An electron-beam lithography machine using a variable-shaped beam is shown schematically in Fig. 1.17. Figures 3.26 and 3.27 show an example of such a machine and examples of the patterns that it can generate.

3.4.3 Limitations

Effects of electron–material and electron–electron interactions should be avoided in the process.

(1) Proximity effect of electrons on material

Scattering of electrons in the resist and backscattering from the wafer cause degradation of geometrical resolution and pattern distortion. This is called the proximity effect.

TABLE 3.5
Examples of electron-beam lithography machines [111]

Scanning method	Stage movement	Beam shape		
		Gaussian round beam	Shaped beam	
			Fixed	Variable
Vector scanning	Step-and-repeat	JBX–5A (JEOL) EBMF (Cambridge Instr.) VS–1 (IBM) EBPG (Philips)	EBSP (TI)	JBX–6A (JEOL) ZBA–10 (Carl Zeiss) EL–3 (IBM) HL–600 (Hitachi)
Raster scanning	Step-and-repeat Continuous	MEBES (Perkin–Elmer) EeBES (Varian) EBM (Toshiba Machine)	EL–1 (IBM)	EL–2 (IBM) VLS–1000 (Varian) AEBLE150 (Perkin–Elmer)

FIG. 3.26. Electron-beam lithography machine (Courtesy Atsugi ECL NTT)

The exposure intensity distribution $G(r)$ (C/cm^2) at a radial distance r from the centre of the projected beam is [112]:

$$G(r) = C_1 \exp[-(r/B_1)^2] + C_2 \exp[-(r/B_2)^2] \quad (3.6)$$

where C_1, C_2 (C/cm^2) and B_1, B_2 (m) are the constants determined from the exposure conditions. The first term is associated with incident primary electrons and the second with backscattered electrons. An experimental result is shown in Fig. 3.28 [112]. In this case, best fit occurs at $B_1 = 0.1$–0.2 μm, $B_2 = 1$–1.2 μm and $C_1/C_2 = 1.5$–3.

This effect can be compensated for by various methods [113]. Computer control of incident electrons [114] makes the processing time longer. A multilayer resist makes resist processing more complicated. A higher acceleration

Fig. 3.27. Examples of patterns generated by electron-beam lithography: FBM-G monolayer resist patterns of field oxide layer after dry etching (Courtesy Atsugi ECL NTT)

voltage [115] increases the penetration range of the electrons larger: thus the effect of incident primary electrons in the resist is reduced, as is that of backscattered electrons, because of the wide scattering. In such a case, a higher deflection voltage is required and more of the electron beam power is wasted in the wafer. A complete method for overcoming the proximity effect has not yet been developed.

(2) Electron–electron interaction

The shaped beam can reduce the necessary exposure time because of its high current, but blurring of the beam edge occurs as a result of electron–electron interactions [105]. This effect can also be reduced by higher acceleration

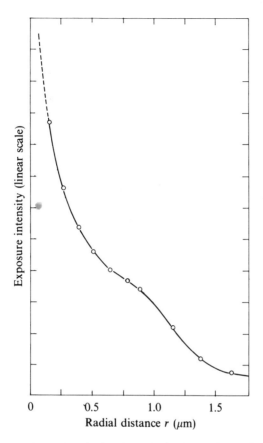

FIG. 3.28. Exposure intensity distribution obtained experimentally using a 25 kV, 0.2 μm diameter incident beam on a silicon substrate with 600 nm PMMA resist developed in MIBK–isopropyl alcohol (1:1) for 60 s [112]

voltages. Therefore, the use of higher voltages appears to be the way to overcome the blurring and proximity effects at the same time. A combination of high current at the centre and low current at the edge of the pattern also seems useful.

3.4.4 Image projection using electron beams

Techniques of mask pattern printing have also been developed using electrons.

(1) Electron image projection by photocathode [116–118]

The principle is shown in Fig. 3.29 [117]. Photoelectrons emitted from a mask coated with a photoemissive layer are accelerated and projected onto the

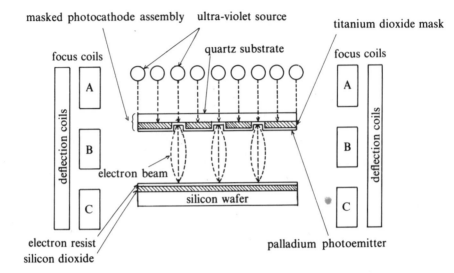

FIG. 3.29. Principle of electron pattern projection by photocathode [117]

wafer by coaxial electric and magnetic fields; thus mask patterns can be transferred to the wafer at a high rate. In this system, the mask structure is analogous to that of photomasks. There is no electron optical cross-over and therefore the image field is flat. However, there are problems such as image distortion, the proximity effect, reimpaction of backscattered electrons, and wafer alignment [119].

(2) Electron image projection by electron beam

The principle is analogous to that of a photo-optical projection system, as shown in Fig. 3.30 [120]. A self-supporting foil mask is irradiated with electrons. The electrons passing through the mask openings are projected onto the wafer with a reduction to 1/10 size. This system can transfer the pattern at a very high printing speed of 1.44×10^6 square elements of 0.25×0.25 μm in 0.1 s.

This transfer method has disadvantages in the production of self-supporting foil masks, such as fracture and distortion of the masks; however, it appears that the mask-making process should be further developed [121].

3.5 Other kinds of electron-beam process

3.5.1 Electron reactive processing

This process is carried out through the formation of radicals by electron bombardment. Its purpose is to change the material properties such as

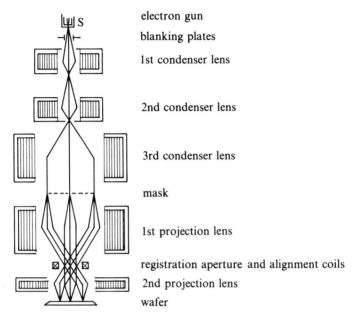

FIG. 3.30. Schematic diagram of electron projection microfabrication system (magnetic lenses not scaled) [120]

resistance to temperature, to solvents, to corrosion or to wear, mechanical properties, dyeability, adhesion ability, surface properties, and sterilization.

The changes in the materials produced by electron-beam irradiation are due to combination or degradation of the molecules. Combination reactions are polymerization (copolymerization), homopolymerization, graft-polymerization, and cross-linking; degradation is represented by depolymerization.

(1) Factors affecting the process

(a) G-value The number of induced reactions is estimated from the irradiating energy. The parameter G is defined as the number of molecules per 100 eV (1.6×10^{-17} J) of absorbed energy. Its value indicates the extent of reaction. Some examples are 1.4–4.6 for polyethylene, 0.6 for poly(vinyl chloride), and 3.2 for natural rubber [122].

(b) Irradiation energy For industrial use, it is convenient to define the irradiation energy required for processing unit mass of material as 1 Gy = 1 J/kg, or 1 rad = 10^{-2} Gy. The material is irradiated with 1 Gy when the radiation rate is 1 J/kg.

(c) Electron energy Thicker materials can be processed by electron beams than by ultraviolet light if the electrons have sufficient energy, because energetic electrons can penetrate deeply into material [123].

(2) Processing equipment

Energetic electrons generated in a high vacuum can pass through a thin foil window of material such as aluminium, titanium, or their alloys, and can be projected onto the material in the external atmospheric environment. There are mainly two types of machine: the scanning beam type and the curtain beam type [124, 125]. The former uses a widely scanned beam, the latter a widely spread beam produced by a line filament electron source. A schematic diagram of a curtain-type machine is shown in Fig. 3.31 [124].

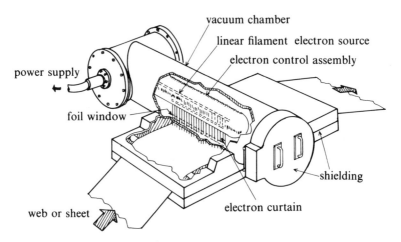

FIG. 3.31. Schematic diagram of curtain-type electron processor [124]

In this system, electron energy is absorbed in the foil window, air gap, workpiece and air behind the workpiece. Relatively high-energy electrons, e.g. 0.3–3 MeV, are required for general use.

(3) Property changes due to combination reactions

Electron irradiation effects depend largely on the kind of material [3, 18, 126]. Mechanical property changes in unsaturated polymer resins are shown in Fig. 3.32 [126]. Young's modulus of the polymer resins increases, but elongation is reduced; furthermore, these are subject to ageing effects. In the figure, the values obtained by the normal catalytic technique are also shown.

The irradiation effect on elongation due to the difference in molecular weight is shown in Fig. 3.33 for polyester diacrylourethane oligomer [127]. Elongation at a molecular weight of 4000 is about 200 per cent, but this is not

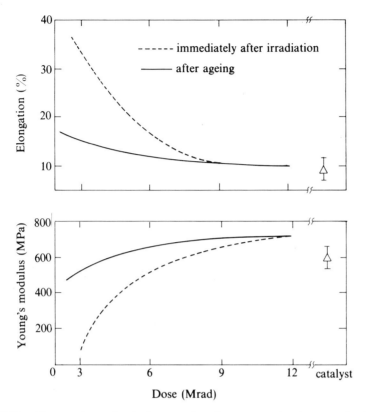

FIG. 3.32. Changes in mechanical properties of unsaturated polymer resin due to electron irradiation [126]

enough for use in textiles. In such a case, addition of a small amount of chain transfer agents is effective in loosening the structure [128].

(4) Property changes due to degradation

The molecular weight of polymers can be reduced by electron irradiation. This effect is called degradation and is used in some fields of industry. For example, cellulose is reduced in molecular weight and its molecular weight distribution is made more uniform. Poly(ethylene oxide) (Polyox) is reduced in molecular weight and becomes soluble in water [129].

(5) Polymerization of residual gases in vacuum

Deposit films are formed on the solid surfaces of electron-beam equipment, owing to polymerization of residual gases by electron-beam bombardment [130, 131]. The polymerization rate depends on pressure and beam current;

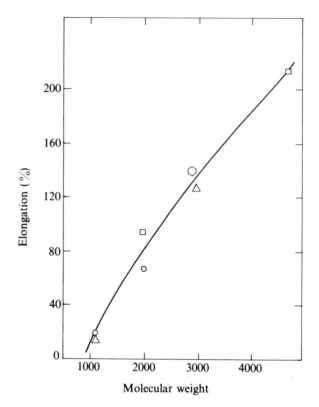

FIG. 3.33. Elongation at break as function of molecular weight of polyester diacrylourethane oligomer (at 5 Mrad) [127]

high pressure and high current cause a high polymerization rate [131]. Furthermore, the epoxy film evaporated on the substrate of glass or cast epoxy and bombarded with electrons has a high electrical insulation capacity [132].

(6) Applications

There are various applications in industry, such as (i) coated wire and cable, and extruded tubings for improving heat, chemical, and wear resistance, (ii) production of polyethylene foam, (iii) production of heat-shrinkable polyethylene film, (iv) vulcanization of rubber, (v) sterilization of medical products, food, and sewage, and (vi) curing of polymer coatings.

Electron-beam curing of polymers has many advantages compared with other curing methods [123, 124, 133–137]: high energy efficiency, high production rate, compact equipment, instant start-up and shut-down, capability of curing thick material, no pollution, etc.

3.5.2 Electron-beam annealing of semiconductors

Semiconductor crystals in which ions are implanted, as described in Chapters 1 and 4, have conventionally been annealed by furnace, because the doping elements must be rearranged in the crystal. Annealing of semiconductors by electron beam has also been carried out [138], and has the advantages that it is feasible to control the annealed depth by electron energy, and easy to control accurately the positioning and motion of the electron beam, and that energy absorption is independent of the surface condition, etc.

There are two types of electron-beam annealing: pulsed and scanning.

(1) Pulsed electron-beam annealing

This process is carried out by a pulsed beam of the order of 10^1 ns duration, obtained by capacitor discharge [139–141]. Data available on the parameters of the electron beam are 1–30 kV, 10–40 kA, 5–7.6 cm diameter, and 50 ns pulse duration [139].

The effects of the process have been elucidated by H^+ backscattering yield, TEM (transmission electron microscopy), SEM (scanning electron microscopy), differential Hall effect, sheet resistivity, etc. [139, 140, 142–145]. The annealing mechanism is considered to be liquid-phase epitaxial regrowth due to melting. The impurity distribution profile after annealing is different from that existing straight after implantation. Figure 3.34 presents arsenic dopant profiles for freshly implanted, furnace-annealed, and pulse-annealed silicon wafers [139]. It is seen that the dopant concentration in pulse-annealed wafers has undergone considerable redistribution.

In thermal annealing of GaAs, the surface must be protected by encapsulants such as Si_3N_4 or SiO_2, or by a suitable ambient pressure of As and Ga, to prevent escape of As and precipitation of Ga [142]. Therefore, transient rapid annealing appears to be a suitable means of avoiding this trouble.

Annealing of Al single crystal implanted with Ga ions has been carried out [145]. This is considered a promising result for the production of metastable alloys with high impurity concentration. The temperature rise of the material is estimated at about 350 K, to a temperature which is below the melting point of aluminium and comparable with the higher temperature applied in thermal treatment.

(2) Scanning electron-beam annealing

This can be carried out using an SEM machine, electron-beam machining, or a welding machine equipped with scanning apparatus. It has the advantages of a constant electron energy and the capability of processing desired portions only. The annealing characteristics and scanning method have been investigated [146–148]. This type of annealing is considered to be solid-phase epitaxial regrowth [146].

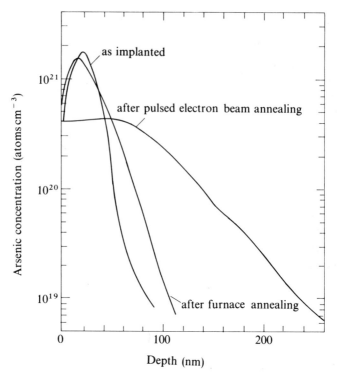

FIG. 3.34. Arsenic impurity profiles in Si $\langle 100 \rangle$ implanted by 25 kV As ion beam (implantation dose $5 \times 10^{15}/cm^2$) [139]

Silicon wafers implanted with As ions can be annealed by an electron beam of 30 kV, 0.5 mA, and 300 μm diameter with a scanning speed of 2.5 cm/s. The electrical activity is the same as that in samples subjected to CW (continuous wave) laser and thermal annealing; moreover, recrystallization is complete [146, 147].

A line source electron beam can reduce the processing time by a factor of 10^{-4}–10^{-6} [149].

This processing technique can also be used for crystal growth of poly-Si film and amorphous Si on the insulating substrate by melting, which is associated with three-dimensional devices [150–152]. Poly-Si film 350 nm thick, deposited on SiO_2, can be annealed by an electron beam of 10 kV, 1.8 mA, and 70–80 μm diameter; the grain sizes obtained are shown in Fig. 3.35 [150], where the dashed lines indicate that the poly-Si film is vaporized. In this process, the beam current, scanning speed, and repetition time are crucial, and the appropriate conditions exist in a rather narrow region [152].

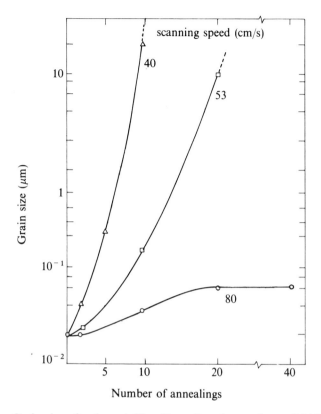

FIG. 3.35. Grain size of polycrystalline films after electron-beam (10 kV, 1.8 mA) annealing as a function of annealing repetition [150]

3.5.3 Electron-beam heat treatment

The solidified layer and heat-affected zone of a workpiece processed by an electron beam are different in structure and hardness from the original materials. This difference is caused by a high cooling rate followed by localized rapid heating of the workpiece by a high power density electron beam.

Heat treatment of a localized portion of a machine part has been developed using a high cooling rate and the excellent controllability of the electron beam. There are two types of method: transformation hardening without melting, and improvement in surface layer properties with melting.

(1) Hardening

This process is carried out by rapid heating and self-quenching. The portion to be hardened is heated rapidly and locally to a point above the transformation temperature but below the melting point, and then rapidly cooled

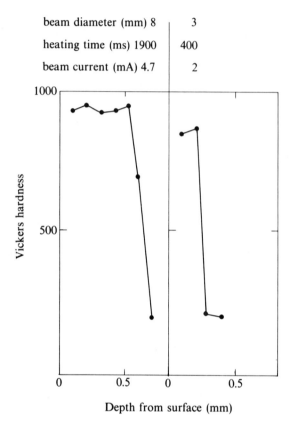

Fig. 3.36. Hardness distribution in steel produced by electron-beam hardening in single pulses [154]

by self-quenching immediately after the electron beam is shut off [153]. The cooling rate is estimated from the CCT (Continuous Cooling Transformation) curves. For steel, the portion to be hardened is heated up to the temperature which produces austenite structure and is then cooled at a sufficiently high rate to produce the martensitic structure.

Figure 3.36 shows the hardness distribution in steel (0.44% C, 0.19% Si, 0.62% Mn, 0.049% P, and 0.030% S) produced by a 150 kV single-pulse electron beam [154]. The hardened depth can be controlled by beam current and pulse duration. The technique based on beam scanning can produce hardening of a required pattern on machine parts [155].

(2) Melting treatment

A surface layer having different properties from the original material is obtained by local melting and subsequent self-quenching. Furthermore, other materials can be coated onto the surface by the electron beam.

(a) Surface melting Improvement of surface properties of materials by this method has been achieved in various kinds of material [156–162]. Iron-based hard materials have shown the most promising results [162]. The relation between scanning speed (400 W electron beam) and hardness obtained is shown in Fig. 3.37 for steel (0.9% C, 1.6% Mn, and 0.25% Si) [161]. A maximum Knoop hardness of about 800 was observed near the bottom at a high scanning speed. The melted zone has in general a highly refined microstructure. Wear tests have been carried out for the same steel, the results of which are listed in Table 3.6 [161]. The electron-beam surface melting increases the hardness and reduces the wear rate.

Corrosion resistances for M2 tool steel processed by electron-beam surface melting have been measured by anodic polarization in borate buffered solution of pH 8.0. Resistance to chemical attack is also improved [162].

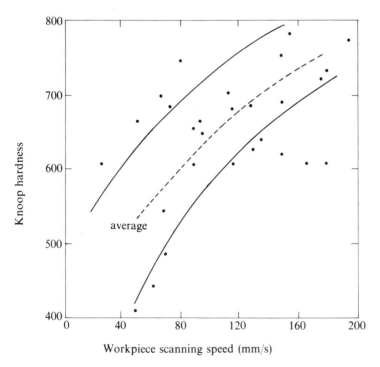

FIG. 3.37. Relation between hardness produced by a 400 W electron beam and scanning speed for steel [161]

TABLE 3.6
Mechanical properties produced in steel by electron-beam surface melting [161]

	Knoop hardness	Hardness ratio	Friction coefficient	Wear rate (mm^3/m)	Wear ratio
Annealed	230	1	0.50	1.6×10^{-4}	1
Hardened	730	0.32	0.74	1.1×10^{-4}	0.69
Electron-beam melting	800	0.29	0.90	0.82×10^{-4}	0.51

Wear rate = wear volume/sliding distance

(b) Modification of surface structure Structures of metals and alloys become amorphous on rapid cooling from the molten state. Cooling rates above 10^{10} K/s are required for the production of amorphous pure metals [163]. For some alloys, the critical rates for obtaining an amorphous layer are quite low. Furthermore, dissimilar alloy structures are produced by coating and melting other kinds of metal deposited on the surface [164–169].

A niobium single-crystal rod 6 mm in diameter was sprayed with $Ni_{60}Nb_{40}$ on the surface and then a scanning electron beam was projected under conditions determined from cooling curves [170]. The hardness distribution obtained is shown in Fig. 3.38 [168].

The relation between cooling rate and hardness has been determined for melted powder alloy (Ni–7Cr–2.8B–4.5Si–3Fe) on steel substrate. The maximum hardness is obtained at a cooling rate of 10^5 K/s [167].

3.5.4 Electron-beam vapour deposition

In this process, atoms or molecules evaporated by the electron beam are used to make a thin film or foil. Evaporation by electrical heating is widely used in conventional vacuum deposition. Evaporation by electron-beam heating has many advantages, such as applicability to high melting point materials and to materials which are reactive to tungsten and tantalum, high energy efficiency because of the direct heating of the material, and highly accurate controllability of evaporation.

(1) Evaporation process

The evaporation rate G (kg/m^{2}s) in a high vacuum depends on the molecular weight of the material M, the temperature T (K), and the saturated vapour pressure p (Pa) and is given [171, 172] by

$$G = 4.4 \times 10^{-3} \alpha p \sqrt{(M/T)} \tag{3.7}$$

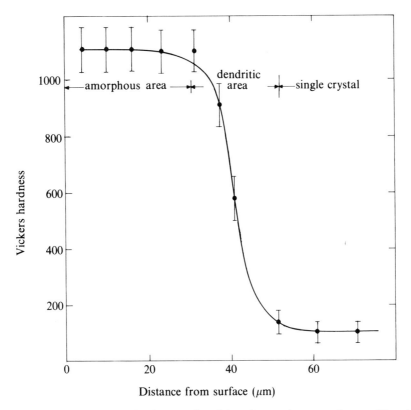

Fig. 3.38. Hardness distribution produced by electron-beam surface melting in $Ni_{60}Nb_{40}$ coated on niobium single crystal [168]

where α is the evaporation coefficient, which depends on the material but is about 1 for most metals. The relation between T and p is given by thermodynamics (Clausius–Clapeyron equation) and the curve for each element is given in Ref. [173]. Once the temperature T is known, the evaporation rate G can be calculated.

The relation between input electron beam power and temperature is complex because the heat losses due to conduction from evaporant to crucible, radiation, and interactions between vapour cloud and electrons must be considered [174].

For alloy evaporant, the temperature of each element is the same; elements with higher vapour pressure are vaporized more rapidly and then the vapour stream has a composition different from that of the alloy. The formation of alloy film should be treated individually in each case.

(2) Electron-beam evaporator

The vapour stream flux density ϕ (kg/m²s) from a point heated source obeys Lambert's cosine law; however, the actual vapour stream produced by an electron beam does not always obey the law. The stream flux density ϕ from a water-cooled crucible is given [175] by

$$\phi = \phi_0 \cos^n \theta \tag{3.8}$$

where θ is the angle from the normal to the vaporizing surface, $n = 2$ to 6, and ϕ_0 is the stream density at $\theta = 0$. To deposit a film with high efficiency, it is necessary to set up the substrate in the direction $\theta = 0$. To prevent the vapour stream entering the electron gun, the gun must be set up off the direction $\theta = 0$. Basic types of evaporator are shown in Fig. 3.39; the beams are deflected largely by magnetic means [176].

The electron-beam gun of the axial evaporator is completely separate from the crucible and the electron beam is bent through 90°. In the transverse evaporator, the gun and crucible are combined into a single unit and the electron beam is bent through 270°. The electron gun of the axial evaporator

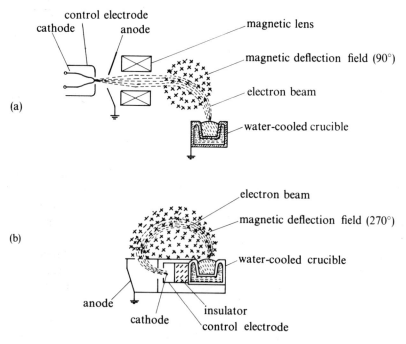

FIG. 3.39. Basic types of electron-beam evaporator. (a) Axial gun. (b) Transverse gun [176]

TABLE 3.7

Evaporation characteristics of various materials irradiated by electron beam [176]

Material	Power (kW)	Evaporation rate (kg/h)	Deposition rate (μm/min)	Specific energy (kWh/kg)
Zr	72	1.17	–	62
Stainless steel	70	4.25	125	16.5
Ti6Al4V	70	2.34	–	30
Inconel	70	2.3	–	30.5
Silica	25	–	150	–
Al	250	46	1200	5

is fairly complex, but is superior in reliability and duration of operation; moreover, the bombardment area can be widened by controlling the bending of the electron beam.

Evaporation rates, deposition rates, and specific energies are listed in Table 3.7 [176].

Although the stream flux density is expressed by eqn. 3.8, a film of uniform thickness on a relatively small substrate can be obtained by rotation and swing of the substrate. For a large substrate, however, it is necessary to use a large-area evaporator. In such a case, the axial evaporator is more useful for sweeping the electron beam. For this purpose, a line evaporator has been developed [176]. Systems for actual production are described in Refs. [178, 179].

Special types of evaporator for research have been developed for ultrahigh-vacuum deposition (10^{-7} Pa or below) [180, 181].

(3) Control of deposited film thickness

Whichever method of thickness measurement is used, films should be deposited to obtain a certain predetermined thickness and required properties. The film thickness is usually monitored during the actual deposition process, to control growth and thickness.

There are various methods of film thickness measurement [182]. Techniques based on the quartz crystal resonator and the ionization gauge are generally used. The former technique detects the change in resonant frequency due to the change in mass deposited on the quartz crystal [183]. The latter detects the ionized vapour. In this case, residual gas molecules in the vacuum are also detected; therefore, the ion current originating from the vapour stream is transformed to an alternating current by a mechanical shutter and thus is distinguished from the current due to the residual gas [184, 185]. Other techniques are based on atomic absorption and electron impact emission spectroscopy [186].

(4) Applications

The process is used for the manufacture of thin-film devices, semiconductors, electric capacitors, etc. in the electronics industry, and for coating glasses, plastics, and mirrors in the optical industry.

The corrosion resistance of steel can be greatly improved by a coating of 1.5 μm aluminium film [179]. Multielement overlay coating of superalloys for gas turbines is superior to diffusion coating [187]. Furthermore, thin foil can be produced by removing the substrate [188–190]. Coatings of refractory metals and ceramics are also applied [191, 192]. The degree of adhesion of the film to the workpiece is evaluated by the pull test [193].

(5) Deposition of alloys

Thin films of alloys can be obtained by this vapour deposition process, the composition of the alloy being the same as that of the vapour stream, which can be controlled. This is done by single or multiple evaporation sources [194]. The vapour pressure of each element, however, must be taken into account when controlling the composition [195]. A15 superconducting compounds may be formed using dual evaporation sources [196]. Aluminium–silicon metallization may also be carried out [197].

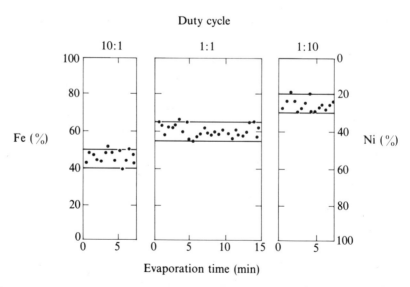

FIG. 3.40. Composition ratios of FeNi in deposited films from dual crucibles [198]

Figure 3.40 shows the changes in composition of FeNi due to the electron-beam sweep in dual crucibles [198]. The composition ratio can be controlled by changing the duty cycle of electron-beam bombardment time in each crucible.

3.5.5 Electron-beam melting

The aim of this process is to purify metals, alloys, and other inorganic materials. In general, electrical induction, resistance heating, and vacuum arc melting processes are used in industry. Compared with these conventional processes, electron-beam processing has the advantages that it may be done in high vacuum and at high temperature, and with high controllability of melting temperature and melting rate; it is therefore considered to be the most suitable method for obtaining high-purity metals, alloys, and other substances. On the other hand, the control of composition with elements of high vapour pressure is difficult.

Practical purification processes are performed by (i) degassing and decomposition of oxides and nitrides, (ii) vaporization removal of impurities in refractory metals, and (iii) removal of metal suboxides, on the basis of their high vapour pressure compared with pure metals [199–203].

(1) Equipment

The ranges of acceleration voltage are 20–30 kV for 60–100 kW, 30–35 kV for 200–300 kW, and 45–50 kV for 1200 kW [204]. One type of equipment is

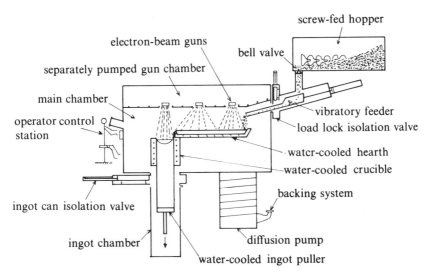

FIG. 3.41. Schematic diagram of electron-beam furnace [205]

shown in Fig. 3.41 [205], in which raw material in the form of pellets or powder is fed onto a hearth and melted there. Other types [206–208] involve (i) feeding raw material in the form of a rod into the electron-beam path and dripping the melted material into the molten pool of the ingot, (ii) feeding the raw material in the form of pellets or powder into the molten pool of the ingot and melting it there, or (iii) feeding the raw material in the form of pellets, powder, or rod onto the hearth and making the ingot there. Equipment with various feeding methods is reviewed in Ref. [209].

The hearth and crucible are generally made of copper with a water cooling jacket, in which melting at temperatures above 3000 K is possible. The puller is moved down according to the growth of the ingot so that the height of the molten pool surface is held constant, as shown in Fig. 3.41.

Industrial plants utilizing this kind of equipment are surveyed in Ref. [210]. Maximum dimensions of the ingot are 600 mm in diameter and 4000 mm in length with 3 MW equipment.

(2) Purification

This process has been developed for the purification of reactive and refractory metals since the early days of electron-beam processing. Recently it has been applied in the production of steel ingots [207] and the recycling of materials such as titanium scrap [205].

Purification and hardness data for reactive and refractory metals obtained by three different vacuum melting techniques are compared in Table 3.8 [211]. Melting by the high-current electron beam of a plasma source will be described in the next section. Changes in metallic solute elements in automotive steel have been determined theoretically and experimentally [212].

Degassing of impurities also depends on the melting time. Removal of some elements in gaseous form and change in hardness of niobium are shown in Fig. 3.42 [213]. Oxygen is degassed in the form of CO. The rate of removal of oxygen is higher than that of carbon [214]. For degassing of the residual carbon in the processing time of 30 min, a specimen with higher oxygen content must be prepared [213].

The mechanical properties and structure of electron-beam-melted refractory metals are also largely different from those of the raw materials [215, 216].

(3) Electron-beam melting using a plasma electron gun

This process is a kind of electron-beam melting, but the electron gun used is a plasma type, in which electrons are extracted from the plasma generated in a hollow cathode as shown in Fig. 3.43 and used for melting [217, 218]. The process is also used for welding because of the high-current beam capability.

Electron beam currents greater than 1000 A can be obtained easily at an acceleration voltage of 100 V. This means that processing can be done

TABLE 3.8
Impurity contents and hardnesses obtained by electron-beam and inert gas arc melting [211]

	Material	Treatment		Impurity (ppm)			Vickers hardness (50 N)
				O	H	C	
Ti	Sponge	—		—	30	430	—
	Button ingot	Inert gas arc melting		1400	36	380	242
		Plasma e.b. melting	A	683	16	330	227
			B	307	24	220	190
			C	290	9	200	159
		High voltage e.b. melting		300	12	210	167
Zr	Sponge	—		930	—	—	—
	Button ingot	Inert gas arc melting		1372	23.0	—	233
		Plasma e.b. melting	A	837	16.2	—	184
			B	827	9.6	—	162
			C	682	10.4	—	134
		High-voltage e.b. melting		695	7.7	—	154
Mo	Scrap	—		121	13.0	—	—
	Button ingot	Inert gas arc melting		72	1.8	—	207
		Plasma e.b. melting	A	79	3.2	—	188
			B	59	4.1	—	154
			C	49	3.2	—	143
		High-voltage e.b. melting		30	0.4	—	151
Ta	Powder	—		430	43	—	—
	Button ingot	Plasma e.b. melting	A	247	5.4	—	98.5
			B	101	1.6	—	93.6
			C	103	2.5	—	65.1
		High-voltage e.b. melting		40	2.8	—	61

Plasma e.b. melting A: Argon for welding
B: High-purity argon
C: Special high-purity argon

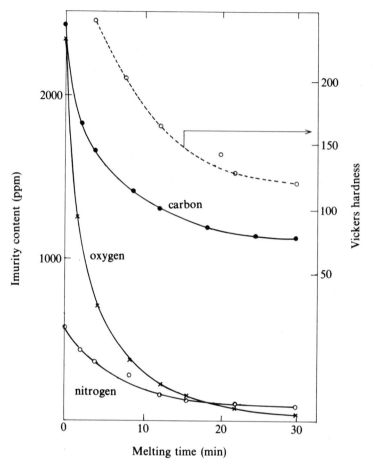

Fig. 3.42. Removal of elements in gaseous form and hardness for niobium in electron-beam melting (10 kV, 320 mA). Chemical composition of workpiece: oxygen 0.235%, carbon 0.241%, nitrogen 0.057% [213]

without X-ray generation and discharge due to gases. Using high-purity argon for plasma generation, degassing effects are nearly the same as those obtained using the ordinary electron gun as given in Table 3.8.

3.5.6 Other processes

(1) Surface polishing

The resolidified molten layer produced by a scanning electron beam generally has a roughened rippled surface. This is caused by liquid flow due to the

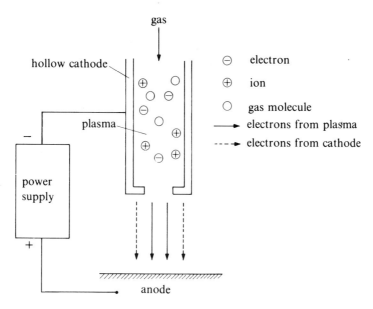

FIG. 3.43. Hollow-cathode plasma gun

surface-tension gradient. This can be avoided by using a scanning speed above a critical value, when the molten layer does not have sufficient time to form ripples [219].

Application of the above technique to thin films in semiconductor devices appears to be difficult in practice. A technique for improving the morphology of as-grown monocrystalline thin films on foreign substrate has been developed using a large-area pulsed electron beam [220]. The surface layer of the thin film is melted, allowing the surface tension to erase irregular features. The crystallinity of the layer is preserved by rapid resolidification, the underlying material acting as a seed.

(2) Bonding of thin foil or wire

Electrons penetrate some distance into solid materials. A thin foil or wire can be welded to the surface of an underlying workpiece, because the latter is heated and melted by electron energy transmitted through the foil or wire. Aluminium foil of 20 μm thickness can be bonded to stainless steel and copper by electron beams of 100 and 120 kV respectively; however, the bonding conditions of the electron beam are very critical [221].

(3) Floating zone melting

Materials of low melting point and low reactivity can be refined in crucibles. However, contamination occurs in melting materials of high melting point and high reactivity. Floating zone melting is useful in avoiding contamination, because the material is heated by high-frequency induction or radiation from a cylindrical refractory metal heater.

Heating by electron-beam bombardment offers advantages in simplicity of operation and also permits the method to be extended to the melting of refractory metals [222, 223]. Processing is done with high energy efficiency compared with other floating zone melting methods. Single-crystal tungsten and molybdenum with excellent bend ductility can be obtained easily [224].

(4) Powder production by melting

Powders of reactive and refractory metals or their alloys may be produced by electron-beam melting techniques [225–227]. This is considered especially useful for titanium alloy powder production.

The basic concept is that metals melted by the electron beam are ejected by centrifugal force, and are atomized in the process. There are two methods for atomization: one-step and two-step [226]. In the one-step method, the ingot to be atomized is clamped at one end and rotated. The free end of the ingot is then melted by the electron beam. The melted portion is thrown off by the centrifugal force and fine particles are formed, which solidify into spherical particles if sufficient cooling takes place before the wall is reached. In two-step atomization, the metals are melted and dripped into a rotating disc or cup. In this method, there are no limitations of shape and size of metal objects to be atomized.

(5) Fine powder production by evaporation

Fine powders may be produced by the evaporation technique in a gaseous environment. The evaporated compounds or elements are cooled and condensed into ultrafine particles in the surrounding gas. Plasma-jet, laser-beam, or electric-current heating are generally used for evaporation. Electron-beam heating has also been applied in this process [228].

Pure metal powder is produced in Ar gas, and nitride powder such as TiN and AlN are produced by evaporating Ti in N_2 gas and Al in NH_3 gas. The particle diameters obtained are less than 10 nm for TiN and 8 nm for AlN [229].

References

[1] Bakish, R., editor, *Introduction to electron beam technology* (John Wiley & Sons, New York, 1962).

[2] Charlesby, A. and Wycherley, V., *Int. J. Appl. Radiation and Isotopes* **2** (1957), 26.

[3] Charlesby, A., *Atomic radiation and polymers* (Pergamon Press, London, 1960).

[4] Sugata, E., editor, *Electron and ion beam handbook* (Nikkan Kogyo Shinbunsha, Tokyo, 1973), 247 (Japanese).
[5] Whiddington, R., *Proc. Roy. Soc. London* **A86** (1912), 360.
[6] ——, ibid. **A89** (1914), 554.
[7] Spencer, L. V., *Phys. Rev.* **98** (1955), 1597.
[8] Archard, G. D., *J. Appl. Phys.* **32** (1961), 1505.
[9] Everhart, T. E. and Hoff, P. H., ibid. **42** (1971), 5837.
[10] Kanaya, K. and Okayama, S., *J. Phys. D, Appl. Phys.* **5** (1972), 43.
[11] Cosslett, V. E. and Thomas, R. N., *Brit. J. Appl. Phys.* **15** (1964), 883.
[12] ——, ibid. **15** (1964), 1283.
[13] ——, ibid. **16** (1965), 779.
[14] Murata, K., Matsukawa, T., and Shimizu, R., *Japan J. Appl. Phys.* **10** (1971), 678.
[15] Shimizu, R., Ikuta, T., and Murata, K., *J. Appl. Phys.* **43** (1972), 4233.
[16] ref. 4, 252.
[17] Tabata, Y., *Hoshasen Jyugo* (Sangyo Tosho, Tokyo, 1969), 6 (Japanese).
[18] Yamazaki, F., editor, *Isotope binran* (Maruzen, Tokyo, 1970), 220 (in Japanese).
[19] Dudek, H. G., *Z. angew. Physik.* **31** (1971), 331.
[20] Miyazaki, T., *J. Appl. Phys.* **48** (1977), 3035.
[21] Merli, P. G. and Rosa, R., *Optik* **58** (1981), 201.
[22] Wells, O. C., *Proc. 3rd Symp. on Electron Beam Technology* (Alloyd Electronics Corp., Boston, Mass., 1961), 291.
[23] Pahlitzsch, G. and Visser, A., *Proc. 3rd Int. Conf. on Electron and Ion Beam Science and Technology* (The Electrochemical Soc., New York, 1968), 335.
[24] Sturans, M. A. and Koste, W. W., *J. Vac. Sci. Technol.* **15** (1978), 1060.
[25] Desilets, B. H., ibid. **15** (1978), 1056.
[26] Pahlitzsch, G. and Kuper, G., *VDI-Z* **111** (1969), 83.
[27] Miyazaki, T., Yoshioka, S., and Kimura, T., *Annals of the CIRP* **25** (1977), 89.
[28] Siekman, J. G. et al., *J. Phys. E. Sci. Instrum.* **8** (1975), 391.
[29] Miyazaki, T. and Kimura, T., *J. Japan Soc. Prec. Engg.* **45** (1978), 854 (Japanese).
[30] Landau, H. G., *Quart. Appl. Math.* **8** (1950), 81.
[31] Pahlitzsch, G. and Visser, A., *Annals of the CIRP* **13** (1966), 147.
[32] Pahlitzsch, G. and Visser, A., *VDI-Z* **109** (1967), 1197.
[33] Dobenek, D. and Parrella, A., *Proc. 6th Int. Conf. on Electron and Ion Beam Science and Technology* (1974), 431.
[34] Dobenek, D. and Steigerwald, K. H., *IEE Conf. Publ.*, **133** (1975).
[35] Closs, W. W. and Drew, J., *SME Paper* **MR78 597** (1978).
[36] Meyer, W. E., *Industrie* **102** (1980), 54.
[37] Miyazaki, T. and Kimura, T., *Bull. Japan Soc. Prec. Engg.* **10** (1976), 35.
[38] Opitz, W. and Steigerwald, K. H., *Schweissen u. Schneiden* **13** (1962), 432.
[39] Steigerwald, K. H., *Feinwerktechnik* **66** (1962), 45.
[40] Arata, Y. and Tomie, M., *Trans. Japan Weld. Soc.* **1** (1970), 2.
[41] ref. 1, 339.
[42] Pahlitzsch, G. and Visser, A., *Annals of the CIRP* **14** (1966), 19.
[43] Schiller, J., Panzer, S., and Heisig, U., *Proc. 6th Int. Conf. on Electron and Ion Beam Science and Technology* (1974) 412.

[44] Schiller, S., Heisig, U., and Panzer, S., *Solid-state technology* (July 1975), 38.
[45] Kuper, G., *VDI-Z* **116** (1974), 401.
[46] Schwarz, H., *J. Appl. Phys.* **35** (1964), 2020.
[47] Miyazaki, T. and Taniguchi, N., *Proc. 5th Int. Conf. on Electron and Ion Beam Science and Technology* (Houston, Texas, 1972) 291.
[48] Tong, H. and Giedt, W. H., *Weld. J.* **49** (1970), 259-S.
[49] Wells, O. C., *Proc. 4th Symp. on Electron Beam Technology* (Boston, Mass., 1962) 105.
[50] Suzuki, H., Hashimoto, T., and Matsuda, F., *J. Japan Weld. Soc.* **32** (1963), 996 (Japanese).
[51] Leskov, G. I., Trunov, E. N., and Zhivaga, L. I., *Avt. Svarka.* **6** (1976), 13.
[52] Lankin, Y. M., *Avt. Svarka.* **2** (1978), 16.
[53] Schauer, D. A., Giedt, W. H., and Shintaku, S. M., *Weld. J.* **57** (1978), 127-S.
[54] Klemens, P. G., *J. Appl. Phys.* **47** (1976), 2165.
[55] Weber, C. W., Funk, E. R., and McMaster, R. C., *Weld. J.* **51** (1972), 90-S.
[56] Taniguchi, N. and Miyazaki, T., *Annals of the CIRP* **21** (1972), 45.
[57] Miyazaki, T. and Taniguchi, N., *Sci. Papers IPCR* **67** (1973), 7.
[58] Arata, Y. and Fujisawa, M., *Proc. 2nd Int. Symp. of Japan Weld. Soc.* (Osaka, 1975), 33.
[59] Tykin, Y. M. and Kuz'min, G. S., *Svar. Proiz.* **9** (1976), 11.
[60] Giedt, W. H. and Wei, P. S., Modelling of casting and welding processes: *Proceedings of 1983 Engineering Foundation Conf.* (New England College, 1983).
[61] Meier, J. W., *Proc. 3rd Symp. on Electron Beam Technology* (Boston, Mass., 1961), 145.
[62] Meier, J. W., *Proc. 2nd Int. Vacuum Congress* (Washington DC, 1961).
[63] Meier, J. W., *Proc. 2nd Int. Conf. on Electron and Ion Beam Science and Technology* (New York, 1966), 505.
[64] Adams, M. J., *Brit. Weld. J.* (March 1968), 145.
[65] Edgar, E. and Dorn, L., *Schweissen u. Schneiden* **20** (1968), 261.
[66] Boncouer, M., Marhic, J. Y., and Rapin, M., *Proc. 3rd Int. Conf. on Electron and Ion Beam Science and Technology* (1968), 358.
[67] Arata, Y., Terai, K., and Matsuda, S., *Trans. JWRI* **2** (1973), 7.
[68] Meier, J. W., *Proc. 1st Int. Conf. on Electron and Ion Beam Science and Technology* (1965), 634.
[69] Paton, B. E. et al.: *Automatic Welding* **25** (1972), 44.
[70] Tsukamoto, S. and Irie, S., *Committee of Electron Beam Welding of Japan Weld. Soc.* **EBW-313-83** (1983) (Japanese).
[71] Russell, J. D., *The Welding Institute Research Bull.* (Dec. 1978), 352.
[72] Hablanian, M. H., *Proc. Electron Beam Symp. 5th Annual Meetings* (Boston, Mass., 1963), 262.
[73] Passoja, D. E., *Brit. Weld. J.* (Jan. 1967), 13.
[74] Reznichenko, V. F. and Erokhin, A. A., *Svar. Proiz.* **3** (1976), 13.
[75] Hashimoto, T. and Matsuda, F., *J. Japan Weld. Soc.* **33** (1964), 726 (Japanese).
[76] Lubin, B. T., *Weld. J.* **47** (1968), 140-S.
[77] Tong, H. and Giedt, W. H., *Trans. Am. Soc. Mech. Engrs., Series C, J. Heat Transfer* **93** (1971), 155.

[78] Swift-Hook, D. T. and Gick, A. E. F., *Weld. J.* **52** (1973), 492-S.
[79] Arata, Y. and Tomie, M., *Proc. 2nd Int. Conf. of Japan Weld. Soc.* (1975), 45.
[80] Miyazaki, T. and Giedt, W. H.: *Int. J. Heat and Mass Transfer* **25** (1982), 807.
[81] Kuroda, H., *Electron beam welding* (Nikkan Kogyo Shinbunsha, Tokyo, 1971), 43 (Japanese).
[82] Bakish, R. and White, S. S., *Handbook of electron beam welding* (John Wiley & Sons, New York, 1964), 147.
[83] Meier, J. W., *Proc. 2nd Int. Vacuum Congress* (Washington DC, 1961).
[84] Suzuki, H., Hashimoto, T., and Matsuda, F., *J. Japan Weld. Soc.* **32** (1963), 618 (Japanese).
[85] Hokanson, H. A. and Meier, J. W., *Weld. J.* **41** (1962), 999.
[86] Arbel, A., *Weld. J.* **41** (1962), 258-S.
[87] Hashimoto, T. and Matsuda, F., *J. Japan Weld. Soc.* **33** (1964), 918 (Japanese).
[88] Arata, Y., Matsuda, F., and Nakata, K., *Trans. JWRI* **1** (1972), 1, and **2** (1973), 1.
[89] Matsuda, F., Hashimoto, T., and Arata, Y., *Trans. Japan Wel. Soc.* (April 1970), 72.
[90] Metzer, G. and Lison, R., *Weld. J.* **55** (1976), 230-S.
[91] Arata, Y., Terai, K., Matsuda, S., and Nakamura, T., *Proc. 2nd Int. Conf. of Japan Weld. Soc.* (1975), 39.
[92] Dietrich, W., *Weld. J.* **57** (1978), 281-S.
[93] Sandstrom, D. J., Buchen, J. F., and Hanks, G. S., *Weld. J.* **49** (1970), 293-S.
[94] ref. 81, 65.
[95] Meleka, A. H., editor, *Electron-beam welding* (McGraw-Hill Pub. Co. Ltd., 1971), 101 (translated into Japanese).
[96] ref. 4, 333.
[97] Sayegh, G., Dumonte, P., and Nakamura, T., *Proc. 2nd Int. Conf. of Japan Weld. Soc.* (1975), 51.
[98] Roudier, R., ibid., 57.
[99] Terai, K. and Nagai, H., *Metal Construction* (Nov. 1978), 536.
[100] Rice, R. W., *NRL Report* (1970), 7085.
[101] Sze, S. M., editor, *VLSI technology* (McGraw-Hill Int. Book Co., 1983), 267.
[102] Ghandhi, S. K., *VLSI fabrication principles* (John Wiley & Sons, Inc., 1983), 534.
[103] Thompson, L. F., *Proc. Tutorial Symp. Semiconductor Technology* (The Electrochem. Soc., 1982), 91.
[104] Hara, T. *et al.*, editors, *VLSI processing handbook* (Science Forum, Tokyo, 1982), 409 (Japanese).
[105] Herriot, D. R., *Proc. Tutorial Symp. Semiconductor Technology* (The Electrochem. Soc., 1982), 139.
[106] Moore, R. D. *et al.*, *J. Vac. Sci. Technol.* **19** (1981), 950.
[107] Edison, J. C. and Scudder, R. K., *J. Vac. Sci. Technol.* **19** (1981), 932.
[108] Moriya, S. *et al.*, *J. Vac. Sci. Technol.* **B1** (1983), 990.
[109] Goto, E. *et al.*, *J. Vac. Sci. Technol.* **B1** (1983), 1289.
[110] Pfeiffer, H. C., *IEEE Trans.* **ED-26** (1979), 663.
[111] Matsumoto, Y., *Proc. Symp. on Semiconductor Devices* (Japan Soc. Prec. Engg., 1984), 27 (Japanese).
[112] Chang, T. H. P., *J. Vac. Sci. Technol.* **12** (1975), 1271.

[113] Matsumoto, Y. and Takigawa, T., *Denshi Zairyo, Kogyochosakai, Tokyo* **22** (1983), 86 (Japanese).
[114] Parikh, M., *J. Vac. Sci. Technol.* **15** (1978), 931.
[115] Yoshimi, M. *et al.*, *Proc. 14th Conf. on Solid State Devices, Tokyo* (Japan Soc. Appl. Phys., 1982), 179.
[116] O'Keeffe, T. W., Vine, J., and Handy, R. M., *Solid-State Electronics* **12** (1969), 841.
[117] Livesay, W. R. and Fritz, R. B., *IEEE Trans. on Electron Devices* **ED-19** (1972), 647.
[118] Livesay, W. R., *J. Vac. Sci. Technol.* **15** (1978), 1022.
[119] Mori, I., *Denshi Zairyo, Kogyochosakai, Tokyo* (Special Issue, 1980), 75 (Japanese).
[120] Heritage, M. B., *J. Vac. Sci. Technol.* **12** (1975), 1135.
[121] Asai, T., *Denshi Zairyo, Kogyochosakai, Tokyo* (Special Issue, 1980), 79 (Japanese).
[122] Tabata, Y., and Araki, K., *Hoshasen Kakoo*, (*Nikkan Kogyo Shinbunsha*, Tokyo, 1969), 55 (Japanese).
[123] Weisman, J., *SME Technical paper* **FC76-489** (1976).
[124] Tripp III, F. P. and Weisman, J., *SME Technical paper* **FC81-425** (1981).
[125] Mizushima, K., *J. Adhesion Society of Japan* **14** (1978), 323 (Japanese).
[126] Charlesby, A. and Wycherley, V., *Int. J. Appl. Radiation and Isotopes* **2** (1957), 26.
[127] Oraby, W. and Walsh, W., *J. Appl. Polymer Science* **23** (1979), 3227.
[128] ——, ibid., 3243.
[129] ref. 122, p. 129.
[130] Christy, R. W., *J. Appl. Phys.* **31** (1960), 1680.
[131] Haller, I. and White, P., *J. Phys. Chem.* **67** (1963), 1784.
[132] Brennemann, A. E. and Gregor, L. V., *J. Electrochem. Soc.* **112** (1965), 1194.
[133] Shiller, S., Heisig, U., and Panzer, S., *Electron beam technology* (John Wiley & Sons, New York, 1982) 493.
[134] Rosetti, L. F., *SME Technical paper* **FC76-510** (1976).
[135] Lauppi, U. V., *Adhesion* **4** (1981), 185.
[136] Bittencourt, E., Ennis, J., and Walsh, W. K., *American Dyestuff Reporter* (Jan. 1978), 32.
[137] Tripp III, E. P. and Nablo, S. V., *J. Coated Fabrics* **8** (1978), 144.
[138] Kirkpatrick, A. R., Minnucci, J. A., and Greenwald, A. C., *IEEE Trans.* **ED-24** (1977), 429.
[139] Greenwald, A. C. *et al.*, *J. Appl. Phys.* **50** (1979), 783.
[140] Itoh, T. *et al.*, *J. Electrochem. Soc., Solid-State Science and Technology* **128** (1981), 2032.
[141] Majini, G. and Nava, F., *Vacuum* **32** (1982), 9.
[142] Tandon, J. L. *et al.*, *Appl. Phys. Lett.* **35** (1979), 867.
[143] Inada, T., Tokunaga, K., and Taka, S., *Appl. Phys. Lett.* **35** (1979), 546.
[144] Davies, D. E. *et al.*, *Appl. Phys. Lett.* **35** (1979), 631.
[145] Hussain, T. *et al.*, *Appl. Phys. Lett.* **37** (1980), 298.
[146] Regolini, J. L. *et al.*, *Appl. Phys. Lett.* **34** (1979), 410.
[147] Ratnakumar, K. N. *et al.*, *Appl. Phys. Lett.* **35** (1979), 463.

[148] Suh, H. T., McMaster, R. A., and Ahmed, H., *J. Vac. Sci. Technol.* **B1** (1983), 827.
[149] Knapp, J. A. and Picraux, S. T., *Appl. Phys. Lett.* **38** (1981), 873.
[150] Shibata, K., Inoue, T., and Takigawa, T., *Appl. Phys. Lett.* **39** (1981), 645.
[151] Knapp, J. K. et al., *Laser and electron-beam interactions with solids*, Appleton, B. R. and Celler, G. K., editors (North-Holland Inc., New York, 1981), 511.
[152] Shibata, K, Inoue, T., and Takigawa, T., *Japanese J. Appl. Phys.* **21** (1982), L294.
[153] Jenkins, J. E., *Metal Progress* (July 1981), 38.
[154] Iwata, A., *Proc. 5th Int. Conf. on Production Engineering, Tokyo* (Japan Soc. Prec. Engg., 1984), 496.
[155] Dreger, D. R., *Machine Design*, (26 Oct. 1978), 89.
[156] Lewis, B. G., Gilbert, D. A., and Strutt, P. R., *Rapid Solidification Processing*, Mehrabian, R., Kear, B. H., and Cohen, M., editors (Claitors Publ., 1980), 221.
[157] Strutt, P. R., Kurup, M., and Gilbert, D. A., ibid, 225.
[158] Mawella, K. J. A. and Honeycombe, R. W. K., *Proc. 4th Int. Conf. on Rapidly Quenched Metals, Sendai, Japan* (1981), 185.
[159] Jenkins, J. E., *Thin solid fabrics* **84** (1981), 341.
[160] Knapp, J. A. and Folistaedt, D. M., *Laser and electron-beam interactions with solids*, Appelton, B. R. and Celler, G. K., editors, (North-Holland Inc., New York, 1981), 407.
[161] Ruff, A. W. and Ives, L. K., *Wear* **75** (1982), 285.
[162] Lewis, B. G. and Strutt, P. R., *J. Metals* (Nov. 1982), 37.
[163] Masumoto, T., editor, *Material science of amorphous metals* (Ohm-sha, 1982), 23 (Japanese).
[164] Grzemba, B. and Ibe, G., *DRAHT* **30** (1979), 288.
[165] Kadalbal, R., Montoya-Cruz, J., and Kattamis, T. Z., *Rapid Solidification Process* (Claitors, 1980), 195.
[166] Tucker, T. R. and Ayers, J. D., ibid., 206.
[167] ——, *Metallurgical Trans. A.* **12A** (1981), 1801.
[168] Bergmann, H. W. and Wordike, B. L., *J. Mater. Sci.* **16** (1981), 863.
[169] Majni, G. et al., *Vacuum* **32** (1982), 11.
[170] Bergmann, H. W., Fritsch, H. U., and Hunger, G., *J. Mater. Sci.* **16** (1981), 1935.
[171] Sawaki, T., *Shinkuu Jochaku*, (Nikkan Kogyo Shinbunsha, Tokyo, 1967), 2 (Japanese).
[172] ref. 4, 351.
[173] Honig, R. E., *RCA Rev.* **23** (1962), 567.
[174] Schiller, S. and Jasch, G., *Proc. 6th Electron and Ion Beam Science and Technology* (1974), 447.
[175] Schiller, S., Heisig, U., and Goedicke, K., *Vakuum Technik* **27** (1978), 75.
[176] ibid., 51.
[177] Schiller, S. and Goedicke, K., *Vakuum Technik* **22** (1973), 149.
[178] Overacker, W. G., *J. Vac. Sci. Technol.* **8** (1971), 357.
[179] Schiller, S., Foerster, H., and Jaesch, G., *J. Vac. Sci. Technol.* **12** (1975), 800.
[180] Andrew, A., *Vacuum* **32** (1982), 376.
[181] Waldrop, J. R. and Grant, R. W., *J. Vac. Sci. Technol.* **A1** (1983), 1553.
[182] Greaves, C., *Vacuum* **20** (1970), 332.
[183] Sauerbrey, G., *Z. Physik* **155** (1959), 206.
[184] Schwarz, H., *Rev. Sci. Instr.* **32** (1961), 194.

[185] ——, *J. Appl. Phys.* **37** (1966), 4341.
[186] Gogol, C. A. and Reagan, S. H., *J. Vac. Sci. Technol.* **A1** (1983), 252.
[187] Boone, D. H., Stragman, T. E., and Wilson, L. W., *J. Vac. Sci. Technol.* **11** (1974), 641.
[188] Smith, H. R., jun., Kennedy, K., and Boericke, E. S., *J. Vac. Sci. Technol.* **7** (1970), S48.
[189] Bunshah, R. F. and Webster, R. T., *J. Vac. Sci. Technol.* **8** (1971), VM95.
[190] Bunshah, R. F. and Juntz, R. S., *J. Vac. Sci. Technol.* **10** (1973), 83.
[191] Bunshah, R. F., *J. Vac. Sci. Technol.* **11** (1974), 814.
[192] Sherman, M. A., Bunshah, R. F., and Beale, H. A., *J. Vac. Sci. Technol.* **12** (1975), 697.
[193] Caulton, M., Sked, W. L., and Wozniak, F. S., *RCA Rev.* **40** (1979), 115.
[194] ref. 133, 169.
[195] Johansson, B. O. et al., *Vacuum* **31** (1981), 247.
[196] Hammond, R. H., *IEEE Trans.* **MAG-11** (1975), 201.
[197] Henger, F. and Feuersrein, A., *Vakuum Technik* **28** (1979), 3.
[198] Schiller, S. et al., *Vakuum Technik* **16** (1967), 207.
[199] Schlatter, R., *J. Metals* **22** (1970), 33.
[200] ref. 4, 295.
[201] Hayashi, C. and Muramatsu, K., *Shinku Yakin*, (Nikkan Kogyo Shinbunsha, Tokyo, 1968, 2nd ed.), 124 (Japanese).
[202] Seagle, S. R., Martin, R. R., and Bertea, O., *J. Metals* **14** (1962), 812.
[203] Aschoff, W. A. and Baroch, E. F., *J. Metals* **14** (1962), 204.
[204] ref. 133, 263.
[205] Landig, T., McKoon, R., and Young, M., *J. Vac. Sci. Technol.* **14** (1977), 808.
[206] Bakish, R., *J. Metals* **14** (1962), 438.
[207] Hunt, C. d' A. and Smith, H. R., *J. Metals* **18** (1966), 570.
[208] Bunshah, R. F., *J. Metals* **15** (1963), 210.
[209] ref. 133, 260.
[210] ref. 133, 277.
[211] Kashu, S. and Hayashi, C., *J. Japan Inst. Metals* **31** (1967), 413 (Japanese).
[212] Andreini, R. J. and Foster, J. S., *J. Vac. Sci. Technol.* **11** (1974), 1055.
[213] Kimura, H., Sasaki, Y., and Uehara, S., *J. Japan Inst. Metals* **29** (1965), 924 (Japanese).
[214] Clough, W. R., editor, *Reactive metals interscience publ.* (New York, 1959), 131.
[215] Takeai, T., *J. Japan Inst. Metals* **28** (1964), 670; **28** (1964), 676; **30** (1966), 1027; **32** (1968), 802 (Japanese).
[216] Fujiwara, T., Katoh, K., and Ohtakara, Y., *ibid.* **41** (1977), 256.
[217] Stauffer, L. H. and Cocca, M. A., *Proc. Electron Beam Symp. 5th Annual Meetings* (1963), 342.
[218] Morley, J. R., *ibid.*, 368.
[219] Anthony, T. R. and Cline, H. E., *J. Appl. Phys.* **48** (1977), 3388.
[220] Tobin, S. P. et al., *Laser and electron-beam interactions with solids*, Appleton, B. R. and Celler, G. K., editors (North-Holland Inc., New York, 1981), 725.
[221] Miyazaki, T., *J. Japan Soc. Prec. Engg.* **43** (1977), 678 (Japanese).
[222] Davis, M., Calverley, A., and Lever, R. F., *J. Appl. Phys.* **27** (1956), 195.
[223] Calverley, A., Davis, M., and Lever, R. F., *J. Sci. Instr.* **34** (1957), 142.

[224] Takaai, T., *J. Japan Inst. Metals* **30** (1966), 1022 (Japanese).
[225] Stephan, H., *AGARD Conf. Proc.* (1976), 200.
[226] Pietsch, W. and Stephan, H., *Science of Sintering* **14** (1982), 35.
[227] Pietsch, W. *et al.*, *Powder Metal. Internat.* **15** (1983), 77.
[228] Iwama, S., Shichi, E., and Sahashi, T., *Japanese J. Appl. Phys.* **12** (1973), 1531.
[229] Iwama, S., Hayakawa, K., and Arizumi, T., *J. Crystal Growth* **56** (1982), 265.

4
ION-BEAM PROCESSING

4.1 Introduction

4.1.1 Concept

In recent years, there has been a growing interest in processing of materials by ion bombardment, because of the increasing practical need for this process in the semiconductor industry.

When an ion impinges on the surface of a specimen and passes through the surface atomic network, it is retarded by elastic scattering due to the atomic nuclei or by inelastic scattering due to electron shells and free electrons. The ion becomes implanted in the target, or may be backscattered by the surface atoms. As a result of the elastic scattering between an ion and a target atom, the latter is displaced from its equilibrium lattice site, and then collides with neighbouring atoms and displaces them in turn. During such collisions, a small number of the displaced atoms are sputtered from the surface of the target. Moreover, in high-energy ion impingement, electrons of the target atom are excited by inelastic scattering during the collision cascade. Subsequent decay of these electrons causes photon emission, and in some cases further electron emission. Thus this process results in the formation of residual ions and a damaged lattice structure in the target, and emission of various kinds of particles such as backscattered ions, sputtered target atoms, secondary and Auger electrons, visible light, or X-rays, as shown in Fig. 1.19 (Chapter 1) and Fig. 4.1 [1].

This process may therefore be applied to the machining or etching of target materials, implantation of various species of atom in the target material, deposition of sputtered atoms on a substrate located opposite the target, and surface component analysis of target materials.

4.1.2 Equipment

Ion-beam processing equipment may be classified into three categories; the focused ion-beam type, the r.f. plasma type, and the ion-shower type. The plasma type and ion-shower type are used mainly for machining or etching, and for deposition of materials. The focused ion-beam type is used mainly for fine machining, ion implantation, surface component analysis, and lithography.

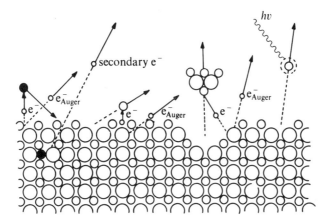

FIG. 4.1. Processes of emission of electrons and photons from a target by ion beam bombardment [1]

(1) R.f. plasma type

A schematic drawing of the high-frequency plasma type of equipment consisting of two parallel electrodes, furnished with a high-frequency oscillator (13.56 MHz), was given in Fig. 1.24 (Chapter 1). The working chamber, made of stainless steel, is evacuated initially to 0.65 mPa using diffusion and rotary pumps, and then argon or some other gas is introduced at a pressure of 6.5 mPa. During the operation, the pressure in the chamber is maintained at 65 mPa by a vacuum pumping system. One of the two electrodes set up in the chamber is a base electrode at earth potential, and the other is a target electrode at a floating potential, to which the r.f. power is applied through a d.c. cut-off capacitor.

The working mechanism of this equipment may be explained as follows. The r.f. power is applied at first to the target and base (substrate) electrodes, causing free electrons in the gaseous state at low pressure between the electrodes to oscillate. The energized electrons bombard the gas atoms, and knock out electrons from the outer shell, thus ionizing the atoms. A plasma, consisting of nearly equal numbers of ions and electrons, is therefore developed between the two electrodes. The ions in the plasma are too heavy to move in response to the r.f. electric field: they remain in the central region between the two electrodes. In contrast, the electrons in the plasma can move easily in response to the alternating electric field.

The base electrode is earthed, but, because the target electrode is cut off from d.c. by the capacitor as shown in Fig. 1.24, the target electrode becomes negatively charged. As a result, ions in the plasma, the potential of which is nearly uniform and earthed, are accelerated in the direction of the target electrode and bombard it at a normal incidence angle. The sputtered atoms from the target are deposited on the substrate of the base electrode mounted opposite the target electrode.

(2) Ion-shower type

Ion-shower equipment can be divided into three major sections: a plasma source, extraction grids, and a working chamber, as shown in Fig. 1.25 (Chapter 1). In the plasma source (ion source), a plasma of argon ions and electrons is generated by electric discharge at a high vacuum (0.1 Pa). That is, the thermionic emission from the heated tungsten wire cathode is accelerated by the anode potential set up in the plasma source. During movement from cathode to anode, the thermionic emission interacts with Ar atoms and ionizes them by electron stripping.

A magnetic field supplied by a coil is applied between the cathode and the anode to cause the electron path to spiral, increasing its effective length and hence increasing the ionization efficiency. Between the plasma source and the working chamber, which is kept at a low vacuum of 10 mPa, are mounted the extraction grids, consisting of arrays of perforated sheets with holes aligned above each other. The outer grid of the plasma source is biased with negative potential (usually at earth potential) with respect to the anode and provides the negative electric field to extract ions from the plasma source. The central grid is biased negatively with respect to earth and prevents electrons from escaping. The inner grid is held at the anode potential. Therefore, only ion flux is extracted from the plasma source and a broad ion beam is formed. Once the ions enter the working chamber, they drift freely to the workpiece holder. To vary the effective incident angle of ions striking the workpiece surface, the workpiece holder is designed to be rotated and swung by electric motors.

For reactive ion-beam etching, ion-shower equipment with an electron cyclotron resonance (ECR) ion source has been developed, as shown in Fig. 4.2 [2].

(3) Focused ion-beam type

A duoplasmatron apparatus specially designed for aspheric lens-making was shown schematically in Fig. 1.23 (Chapter 1). As shown in this figure, Ar or another inert gas is first introduced into the ion source chamber, maintained at a vacuum of 1 Pa through a gas inlet. It is ionized by thermionic emission from a heated tungsten wire cathode and accelerated by the anode. Then arc discharge occurs between the cathode and the anode. The plasma flows into the ion acceleration chamber through the throttle holes of 1.0 mm diameter in the intermediate electrode and the anode. A magnetic field is applied by an electromagnet to contract the plasma flow. The intermediate electrode and the anode consist of soft iron, since these parts are the components of the magnetic circuit. The exciting magnetic coil, the intermediate electrode, the

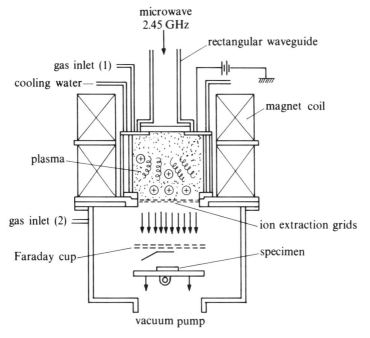

FIG. 4.2. Diagram of ion-shower equipment with ECR ion source (ion beam diameter 15 cm) [2]

anode, etc. are usually water-cooled. A high positive voltage with respect to earth is applied to the ion source chamber, in which the cathode, the intermediate electrode, and the anode are electrically insulated from one another. Through this ion source chamber, a narrow ion beam of high current density is extracted by the pull-out and acceleration electrodes (with negative potential relative to ion source chamber), and focused on the workpiece surface by the electrostatic lens.

In general, this type of equipment has deflection electrodes to scan the ion beam in the x and y directions, and also a workpiece holder which can be swung and rotated by electric motors to vary the machining position.

For the ion-beam machining of electric insulators, a neutralizing ion charge must be supplied by an electron injection shower, because the surface of electric insulators carries a potential repulsive to the incident ions. Therefore, ion-shower and duoplasmatron equipment must have an ion neutralizer to provide thermionic emission from heated tungsten wire immersed in the ion beam.

Recently, equipment of the focused ion-beam type with a liquid ion source, e.g. Ga, has been developed in many laboratories [3, 4]. In this apparatus, a fine (nearly 1 μm diameter) micro-ion beam of high current density and brightness can scan the workpiece by means of a controlling electrostatic lens. Using this type of apparatus, maskless etching of line patterns of sub-micrometre size can be performed. Moreover, for the production of silicon semiconductor devices, an apparatus generating B, Ga, As, and Si ions has been developed [3].

4.2 Ion-beam removal

Usually, micro-machining of precise mechanical parts and micro-etching of electronic devices have been performed by chemical etching. However, plasma etching has been developed for delineating patterns on solid-state devices, because of the complexity and hazards of chemical etching and the necessity to reduce the line-width of the pattern and to automate the process. In a general way, etching occurs in a reactive plasma when material to be etched produces volatile compounds by reaction with active species, as produced in a plasma. In the absence of a significant voltage, ions and neutral radicals in the plasma react with the material at random angles of incidence and produce isotropic etching. As a result, this kind of isotropic plasma etching can lead to a substantial reduction in line-width or the loss of the etched pattern, i.e. undercutting. Moreover, a sufficient etching rate cannot be obtained for SiO_2 and aluminium.

Therefore, reactive ion etching has recently been developed for the pattern delineation of VLSI instead of ordinary plasma etching. More recently, reactive ion-beam etching has been studied in order to understand the mechanism of reactive ion etching. Reactive ion-beam etching is very useful for the pattern delineation of VLSI. Furthermore, ion-beam-assisted chemical etching, which is accomplished by impinging ions into a chemically reactive gas, has been developed for enhanced and localized chemical etching. Ion-beam etching using a reactive gas seems to be the most appropriate dry etching technique for the pattern delineation of VLSI. Types of dry etching technique using ion beams are summarized in Table 4.1 [5].

4.2.1 Ion-beam sputter machining or etching

Ion-beam sputter machining or etching is based on the sputtering of a workpiece or a target bombarded by energized ions of 1 to 20 keV, and is often called ion polishing or ion milling. The sputtering is basically the knocking of surface atoms out of the target due to momentum transfer from impinging ions. In other words, an impinging ion makes sequential elastic collisions with target atoms and knocks them out from the surface layer. This process is therefore a physical phenomenon, or a non-thermal, dynamic

TABLE 4.1
Types of dry etching process [5]

Process	Barrel etch	Planar PE	Planar RIE	CAIBE	RIBE	Ion milling
Control over ion energy	None	Semi-independent	Semi-independent	Independent	Independent	Independent
Control over neutral flux	Independent	Semi-independent	Semi-independent	Independent	Semi-independent	Semi-independent
Control over ion flux		Semi-independent	Semi-independent	Independent	Independent	Independent
Selectivity	Excellent	Very good	Good	Good	Good	Poor
Profile	Isotropic	Often isotropic	Often anisotropic	Isotropic to anisotropic	Anisotropic	Anisotropic
Degree of profile control	None	High	High	High	Low	None
Mechanisms	Chemical	Chemical	Physical + chemical	Physical + chemical	Physical + chemical	Physical

PE = plasma etching, RIE = reactive ion etching, CAIBE = chemically assisted ion-beam etching, RIBE = reactive ion-beam etching process. This atomic-scale stock removal process can be applied to hard and brittle materials such as diamond and ceramics, and to the manufacture of ultra-fine precision parts of electronic, optical, and mechanical equipment.

Several machining characteristics of the process are described in the following sections.

(1) Sputtering yield

The sputtering yield S (atoms/ion) is defined as the mean number of atoms removed from the surface of a target per projected incident ion. This is the most important machining characteristic of ion-beam sputter machining (ion sputter machining) and ion-beam sputter deposition (ion sputter deposition). The sputtering yield S depends on the ion energy, the mass of the incident ion, the mass of the target atom, the ion incident angle to the target surface, the crystallinity and the crystalline orientation of the target material, the temperature of the target during machining, and the partial pressure of oxygen in the residual gas. The dependence of the sputtering yield S on incident angle and ion energy have been investigated, mainly experimentally, but partly theoretically.

The sputtering yield S of amorphous or polycrystalline materials have been predicted theoretically using the macroscopic statistical theory of Thompson [6]. Sigmund [7], and Brand et al. [8], which uses the assumption of random retardation of ions and displaced target atoms in an infinite medium. The sputtering yield S predicted by Sigmund's theory at an ion incidence angle ranging from 0° to 60°, as a function of ion incidence angle θ and ion energy E, is given by

$$S(E, \theta) = \frac{1}{\cos \theta} \frac{0.042}{U} \alpha(M_i/M_t) \frac{1}{N} \left(\frac{dE}{dx}\right)_n \tag{4.1}$$

where M_i is the mass of the ion (kg), M_t is the mass of the target atom (kg), U is the sublimation energy of the target (eV), and N is the number of atoms in thickness dx (atoms/nm). The term $\alpha(M_i/M_t)$ represents the efficiency of energy transfer in elastic collision, and $(dE/dx)_n$ (eV/nm) represents the nuclear stopping power between the ion and the target atoms for the Thomas–Fermi potential. These terms are shown in Figs. 4.3 and 4.4 [9].

Experimental data on sputtering yield as a function of ion energy are shown in Fig. 4.5 [10] together with the curve obtained by Sigmund's theory. As can be seen, the experimental results agree with the predictions of eqn. (4.1). The sputtering yield increases with the incident ion energy and reaches a broad maximum in the energy region 5 to 50 keV. The decrease in sputtering yield beyond 50 keV is due to the deeper penetration of the ions into the target and lower energy release in the surface layer of the target (i.e. the incident ion penetrates the surface layer without any atomic collision).

The dependence of sputtering yield on ion energy for a polycrystalline

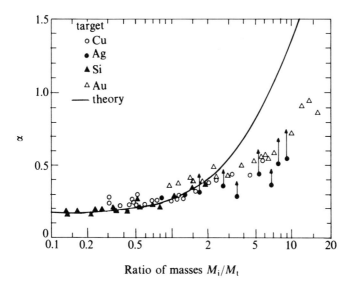

FIG. 4.3. Variation of α at normal ion incidence with the ratio of masses M_i/M_t [9]

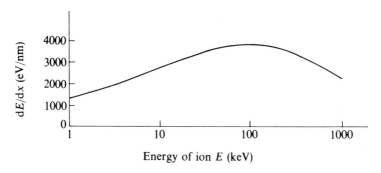

FIG. 4.4. Calculated nuclear stopping power dE/dx for Kr–Cu (Lindhard *et al.*) [9]

copper target sputtered by Xe, Kr, Ar, Ne, and N ions is shown in Fig. 4.6 [11]. This shows that the sputtering yield increases with increasing mass of the ion species. The dependence of sputtering yield S on ion species can be explained intuitively by theory based on kinetic momentum transfer between incident ion and target atoms. That is, when an ion strikes target atoms at rest, the energy E_s (J/atom) transferred from the ion to the target atoms in head-on collision is given by

$$E_s = 4EM_iM_t/(M_i+M_t)^2 \qquad (4.2)$$

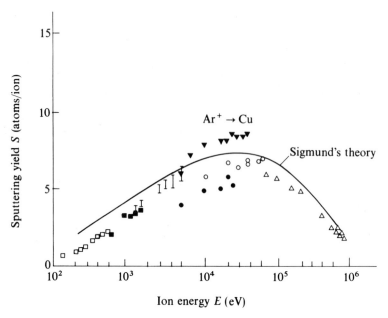

FIG. 4.5. Variation of sputtering yield S with energy E of bombarding ion [10]

where E is the kinetic energy of the incident ion (J/atom), and M_i, M_t are as defined for eqn. (4.1).

Considering eqn. (4.2), in the case where $M_i = M_t$, all the kinetic energy of the incident ion is transferred to the target atoms at rest. Therefore, the closer the mass of the incident ion is to the mass of the target atom, the more efficiently the kinetic energy of the former is transferred to the latter. However, the sputtering yield depends not only on the efficiency of kinetic energy transfer from the ion to the target atom (as shown in eqn. (4.1)), but also on the nuclear stopping cross-section between them (as shown in eqn. (4.2)). In general, an ion of higher atomic number has a larger nuclear stopping cross-sectional area. Therefore, as shown in Fig. 4.6, the sputtering yields S of a copper target sputter-machined by Kr or Ar ions, whose masses are close to the mass of the target atom, are not necessarily the highest. In fact, the sputtering yield is highest for a target sputter-machined with Xe ions, the atomic number of which is greater.

The sputtering yield S depends on the direction of ion incidence relative to the target surface. An example of the angular dependence of the sputtering yield is shown in Fig. 4.7 [12]. As can be seen, the sputtering yield S increases monotonically with ion incidence angle to a maximum at 60° or above, and then decreases rapidly. The incidence angle at the maximum sputtering yield increases with the bombarding ion energy and its mass.

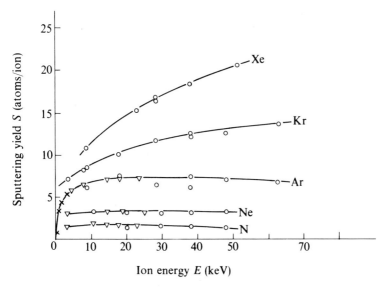

FIG. 4.6. Variation of sputtering yield with mass and energy of ion for a copper target [11]

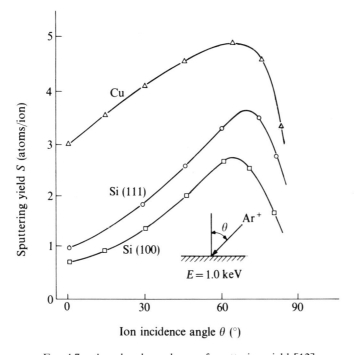

FIG. 4.7. Angular dependence of sputtering yield [12]

The reason why sputtering yield increases and then decreases as the incidence angle increases is considered to be as follows. The amount of energy transferred from ion to target atoms in the top two or three atomic layers increases with increasing incidence angle. Consequently, the target atoms are more easily knocked out. However, when the ion incidence angle exceeds a certain value, a larger number of ions are merely reflected at the target surface, without sputtering any atoms.

The angular dependence of sputter machining rate $V_n(\theta)$ (nm/min mA) of various materials is shown in Fig. 4.8 [13]. The trend of increase and decrease in sputter machining rate varies with the target material. In general, for materials of higher atomic number the curve has a flat section in the range of ion incidence angle 0° to 45°.

The sputtering yields of silver, copper, and tantalum as a function of atomic number of bombarding ion are shown in Fig. 4.9 [14]. The sputtering yield shows a periodicity with respect to atomic number similar to that observed for

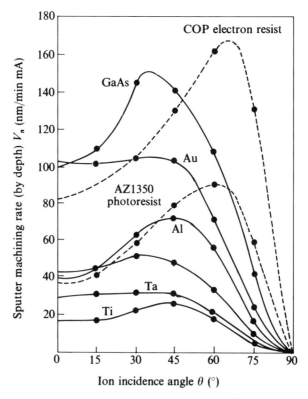

FIG. 4.8. Angular dependence of sputter machining rate V_n. $E = 500$ eV (Ar$^+$ ion) [13]

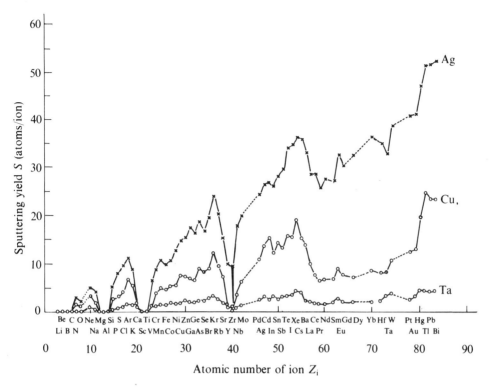

FIG. 4.9. Variation of sputtering yield with atomic number of bombarding ion. $E = 45$ keV [89]

its dependence on the target material. The maximum sputtering yield is reached for each row in the periodic table as the electron shells are filled, i.e. for the inert gas ions. Moreover, the sputtering yield as a function of the atomic number of the target or of the bombarding ions presents the same periodicity as does the surface bonding energy or the inverse sublimation energy of the target.

For single-crystal materials, the sputtering yield depends also on the crystalline structure and the incident direction of the ion beam relative to the crystal orientation, as show in Fig. 4.10 [15]. For ion bombardment close to the three major close-packed crystal axes, the yields are lower by a factor of about 2 to 5 than for any other direction of ion incidence. This is because the probability that collision will create primary recoil atoms, as well as the development of collision cascades, is influenced by the crystal structure, owing to channelling, blocking, or focusing effects. In general, the decrease in sputtering yield for ion-beam incidence along the close-packed crystal direction can be explained by a channelling effect.

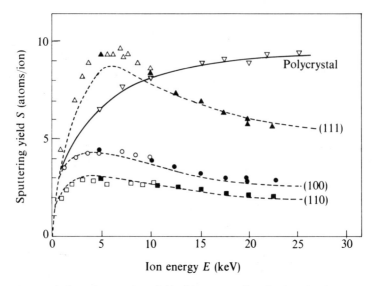

FIG. 4.10. Variation of sputtering yield with energy of ion for bombardment on (111), (100), and (110) surfaces of copper by Ar^+ ions [15]

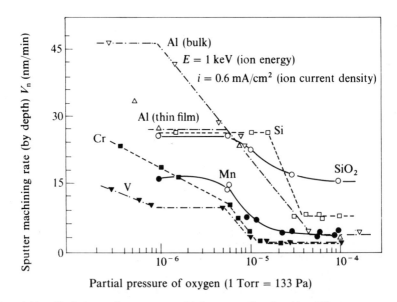

FIG. 4.11. Variation of sputter machining rate (by depth) with oxygen partial pressure. $E = 1$ keV; ion current density $= 0.6$ mA/cm^2 [16]

The effect of partial pressure of oxygen gas in the working chamber on the sputter machining rate for various materials is shown in Fig. 4.11 [16]. The sputter machining rates of metals such as Ti, Cr, and Al decrease with increasing partial pressure, owing to the formation of a chemisorbed layer of oxygen on the active metal surface. However, the partial pressure of oxygen has little effect on the noble metals such as Pt and Au, or on oxides such as SiO_2 and Al_2O_3.

(2) Specific sputter machining rate (by depth)

To describe the sputter machining characteristics in a more practical form, the concept of specific sputter machining rate V_s is introduced. This is defined as the ratio of the depth removal rate to the ion-beam current density:

$$V_s = \frac{(d/\tau)}{(I/a)}$$

where d (cm), I (A) and a (cm^2) are as defined in Fig. 4.12 and τ is the machining time (s). The sputtering yield S (atoms/ion) is given by

$$S = \left(Ad\rho \frac{NZ_t}{M}\right) \Big/ \left(\frac{I\tau}{\varepsilon}\right)$$

where A (cm^2) is as defined in Fig. 4.12, ρ is the density of the target material (g/cm^3), N is Avogadro's number, Z_t is the atomic number of the target material, M is its molecular weight, and ε is the elementary charge (C/ion). Since $a = A\cos\theta$, the specific sputter machining rate is related to the

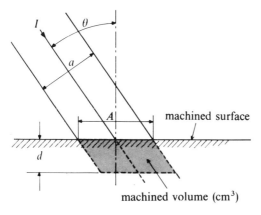

FIG. 4.12. Definition of specific sputter-machining rate. I = ion beam current; a = cross-sectional area of ion beam; A = ion projected area (cm^2); d = machined depth

sputtering yield by the following equation:

$$V_s = \frac{SM \cos \theta}{\rho \varepsilon N Z_t} \tag{4.3}$$

Hence the sputtering yield S can be calculated if V_s is determined experimentally.

A sputter machining rate S_e (cm^3/C) can also be defined, as

$$S_e = SM/NZ_t \rho \varepsilon$$

and $V_s = S_e \cos \theta$.

(3) Angular and energy distribution of sputtered atoms

The angular distribution of atoms sputtered from the target for different primary ion energies, in the case of normal incidence, is described to a first approximation by the cosine law as shown in Fig. 4.13 [17]. For heavy ions and lower ion energies close to the threshold energy of sputtering, more atoms are emitted at larger angles, while for light ions and higher ion energies, more atoms leave the target surface in the perpendicular direction. For oblique ion incidence, the maximum of the emission distribution is shifted a little from the direction of the incoming ion beam (some examples are shown in Appendix B).

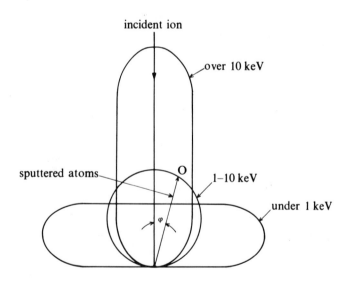

FIG. 4.13. Angular distribution of sputtered atoms from a point P for different primary ion energies. The number of sputtered atoms in a given direction is proportional to the length of the vector PO [17]

The energy distribution of the atoms sputtered from the target is shown in Fig. 4.14 [18]. These atoms have a high average kinetic energy, in the region of 10 to 40 eV. The location of the maximum in the energy distribution does not depend on the bombarding ion energy, but the population of the high-energy tail of the distribution increases with increasing bombarding ion energy. The average energy of the sputtered atoms ceases to increase at ion energies above about 1 keV.

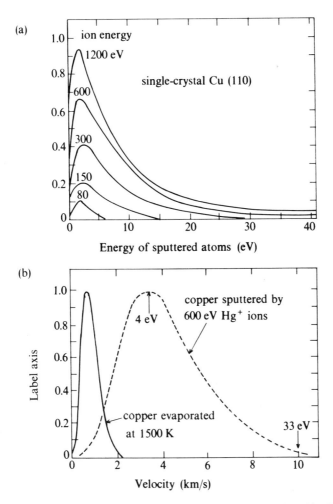

FIG. 4.14. (a) Energy distribution of sputtered copper atoms produced by Kr ions at various bombarding energies. (b) Comparison of velocity distribution of sputtered and evaporated atoms [18]

Special features of ion-beam sputter machining or etching may be summarized as follows:

(i) Ion-beam sputter machining is carried out in vacuum, using inert gas ions which do not react chemically with the atoms of the target, so it is a chemically 'clean' process.
(ii) As the process is insensitive to the chemical properties of the target atoms, hard and brittle materials can easily be machined.
(iii) Machining conditions, such as positioning, speed of travel, current density, acceleration voltage, etc., of the ion beam can easily be controlled electrically.
(iv) The machined profile and dimension of the workpiece could in principle be accurately obtained by means of electrical beam control systems, but at present, a mask is necessary for machining.
(v) Anisotropic machining can easily be performed.
(vi) Ion-beam sputter machining is a non-thermal process based on an elastic collision sequence.
(vii) Ion-beam sputter machining can be done without any residual strain in the workpiece surface.
(viii) The ion beam can be considered to be a wear- and deformation-free tool, in contrast to ordinary solid tools and abrasives.

4.2.2 Reactive ion-beam etching

Ion sputter machining by a beam of chemically inert Ar ions has the inherent characteristics of facet and trench formation and redeposition of sputtered atoms to the target surface, but poor selective processing between target and film or resist, and poor machining or etching rate. However, special advantages of the process are the possibility of high resolution and anisotropic processing. Reactive ion etching has the advantages of anisotropic etching, high etching rate and fairly good selectivity, but the disadvantages of poor end-point detection, poor reproducibility and poor control over the critical process parameters and contamination. For these reasons, reactive ion-beam etching has been developed, which combines the major advantages of reactive ion etching (processing selectivity and high etching rate) with those of ion-beam machining (high resolution and anisotropic processing). In reactive ion-beam etching, reactive gases such as CF_4 and CCl_4 are introduced into the ion source of an ion-shower type of sputter machining apparatus, and ionized by d.c. or microwave discharge. The reactive ions are then extracted from the ion source and delivered to the target. Etching seems to be effected by physical sputtering with reactive ions, chemical reaction between the target surface atoms and the reactive ions, and reactive ion-beam-assisted chemical reaction between the reactive gas and the target surface atoms.

Special features of reactive ion-beam etching are as follows:

(i) Anisotropic etching can be done, then ultra-fine patterning of micrometre feature size can easily be performed.
(ii) A high etching rate and good etching selectivity between the target and film or resist are obtainable.
(iii) Residue-free etching can be carried out.
(iv) Post-etch corrosion is reduced.
(v) Physical and electrical isolation of the target from the plasma yields excellent etching uniformity while minimizing radiation damage.
(vi) Independent controls on parameters of the process can optimize the etching.
(vii) Directionality of the beamed reactive ions results in high anisotropic etching capability.
(viii) Electrical detection of the process end point is possible, which provides excellent stopping capability for the underlying substrate.

(1) Etching rate vs. ion energy

In ion-beam sputter machining, the machining characteristics are represented by the sputtering yield S and specific sputter machining rate by depth, V_s. However in reactive ion-beam etching, the machining rate is not proportional to the ion beam current, so the etching rate V_e (μm/h) is used to represent the machining characteristics. The effects of ion energy, ion incidence angle, reactive gas pressure, and ion current density on the etching rate and selectivity of processing are described below.

The energy dependence of the etching rate and etching selectivity has been investigated for C_4F_8–SiO_2 [19], $SiCl_4$–(SiO_2, Si) [19], CF_4–(SiO_2, Si) [20], C_2F_6–(SiO_2, Si) [20], C_3F_8–(SiO_2, Si) [20], CCl_4–GaAs [21], and Cl_2–InP [22] systems. Examples of the energy dependence of the etching rate are given in Figs. 4.15 [19] and 4.16 [20]. As can be seen, the etching rate increases with ion energy, but reaches a saturation level at ion energies above 1.5 keV. The relation between ion energy and etching rate V_e is given by

$$V_e = V_{e0} \exp(-E_0/E) \tag{4.4}$$

where E is the ion energy and E_0 is a constant corresponding to the activation energy for sputter removal from the workpiece. The constant V_{e0} is determined by ion current density and ion composition. This equation is analogous to the Arrhenius reaction rate formula (see eqn. (1.13) in Chapter 1), with E in place of the thermal kinetic energy kT. The solid line in Fig. 4.15 shows the curve calculated on the assumption that $V_{e0} = 75$ nm/min and $E_0 = 450$ eV. The ion energy E in the etching reaction may be regarded as similar to the thermal energy in chemical reaction. However, for other materials such as Si, the removal of carbon accumulated on the substrate surface must be taken into account.

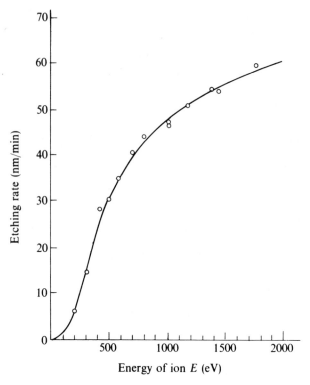

FIG. 4.15. Dependence of SiO_2 etching rate on ion energy [19]. Calculated result by $V = V_0 \exp(-E_0/E)$, where $E_0 = 450$ eV, $V_0 = 75$ nm/min. Gas = C_4F_8; Ion current density = 0.2 mA/cm²

As shown in Fig. 4.16, the etching rate of SiO_2 in the fluorocarbon (CF_4) ion beam increases almost linearly with ion energy and is several times that achieved with Ar ions, while the etching rate of Si is lower than that obtained with Ar ions at ion energies below 800 eV but then rises linearly with ion energy to a value roughly twice that found in Ar ion-beam sputter etching at 1200 eV. The increase in the etching rate of SiO_2 by fluorocarbon ion beams above that of Ar ion-beam sputter etching has been considered as an enhancement of etching due to the formation of volatile SiF_4 by chemical reaction in conjunction with ion bombardment. The suppression of the etching rate of Si in fluorocarbon at ion energies less than 800 eV is due to the accumulation of carbon deposited from the fluorocarbon ion beam on the Si surface, while the rise in etching rate above 800 eV has been attributed to the removal of the protective carbon layer by physical sputtering by the reactive ions. However, the change in the slope of the Si etching curve is not well understood [20].

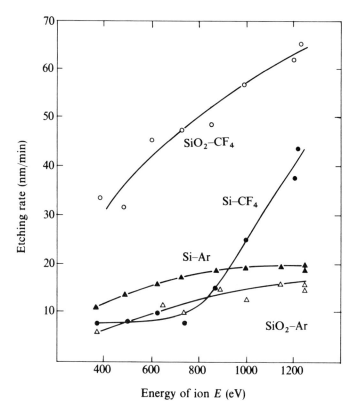

FIG. 4.16. Etching rate vs. ion energy using Ar and CF_4 for SiO_2 and poly-Si [20]. Ion current density = 0.4 mA/cm^2; gas pressure in chamber = 10^{-4} torr, flow rate = 1 cm^3/min for CF_4, and 3 cm^3/min for Ar; cryopumped system

(2) Etching rate vs. ion current density

Relationships between the ion current density and the machining or etching rate have been investigated for C_4F_8–(SiO_2, Si) [19], C_2F_6–(SiO_2, Si) [20], and Cl_2–InP [22]. One example is shown in Fig. 4.17 [19]. It is evident that the etching rate of SiO_2 increases almost linearly with ion current density. However, in the same case, saturation of the etching rate of SiO_2 at ion current densities over 0.3 mA/cm^2 was reported. Furthermore, the etching selectivity (ratio of etching rates of film and substrate) for SiO_2 vs. Si increases with increasing ion current density, and approaches 20.

(3) Etching rate vs. ion incidence angle

The angular dependence of the etching rate has been investigated for CHF_3–SiO_2 [23], CF_4–$LiNbO_3$ [24], CHF_3–$LiNbO_3$ [24], Cl_2–InP [22],

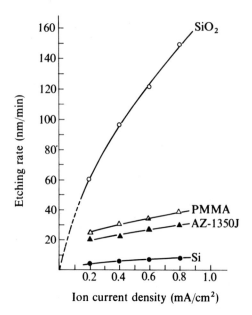

FIG. 4.17. Relation between ion current density and etching rate of SiO_2, Si, PMMA (poly(methyl methacrylate)) and AZ-1350J (positive-type photoresist) [19]. Energy of ion = 1000 eV; C_4F_8 gas pressure = 8×10^{-4} torr

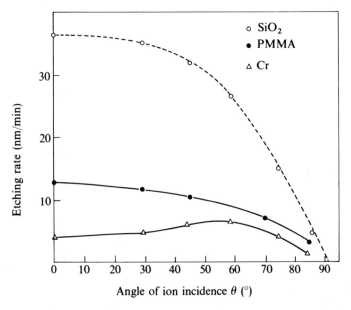

FIG. 4.18. Angular dependence of etching rate of SiO_2, PMMA and Cr for CHF_3 gas at 8×10^{-5} torr [23]. Ion energy = 500 eV; ion current density = 0.4 mA/cm^2

and Ar/Cl$_2$–Si [25]. As shown in Fig. 4.18 [23], the etching rates of SiO$_2$ and PMMA decrease gradually with increasing ion incidence angle. The angular dependences for SiO$_2$ and PMMA etched with reactive CHF$_3$ gas differ from those for Ar ion-beam sputter etching, because the mechanism of reactive ion-beam etching using CHF$_3$ gas includes both physical sputtering and chemical reactive etching, while that of ion-beam etching using Ar gas is only physical sputtering. On the other hand, the angular dependence for non-reactive Cr is independent of the gas species used.

(4) Etching rate vs. gas pressure

The effects of reactive gas pressure on the etching rate have been reported for SiO$_2$ with CF$_4$ [26], and Si and SiO$_2$ with C$_2$F$_6$ [27], as well as the effects of O$_2$ [27–29] or H$_2$ [28] partial pressure in the reactive gas for various materials. Examples are shown in Figs. 4.19 [27] and 4.20 [28], respectively.

As shown in Fig. 4.19, the etching rate reaches saturation at reactive gas pressures greater than 10^{-4} torr (~ 10 mPa) because the ion current density of CF$_3^+$ saturates at a higher reactive gas pressure. Moreover, as shown in Fig. 4.20, the etching rate of SiC with CF$_4$ gas becomes a maximum when the mixing ratio of O$_2$ gas is 40%. This is because the carbon layer on the SiC

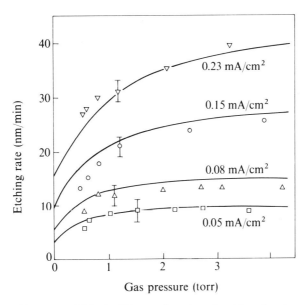

FIG. 4.19. Etching rate of SiO$_2$ in CF$_4$ as a function of etching chamber gas pressure at various ion beam current densities [90]. Ion energy = 500 eV; angle of ion incidence = 30°

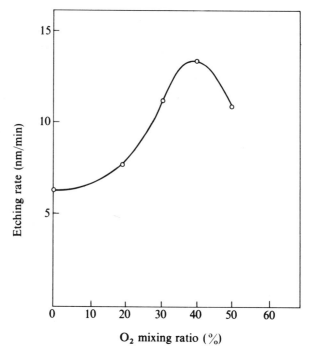

FIG. 4.20. Relation between O_2 mixing ratio in CF_4 gas and SiC etching rate [28]. Gas pressure = 8×10^{-3} torr; ion energy = 500 eV; ion current density = 0.4 mA/cm^2

surface reacts with oxygen, forming volatile CO and CO_2, so providing a clean surface:

$$SiC + CF_n(ion) + O_2 \rightarrow SiF_4 + CO, CO_2 \qquad (4.5)$$

(5) Etched depth vs. etching time

The etching time dependence of the etched depth of $LiNbO_3$ with CHF_3 gas is shown in Fig. 4.21 [24], which indicates that the etched depth is proportional to the etching time. Therefore, the endpoint detection used in reactive ion etching is no longer necessary, because the etched depth can be controlled by etching time, and good reproducibility can be achieved.

4.2.3 Ion-beam-assisted chemical etching (IBAE)

Ion-beam-assisted chemical etching [30], in which the several chemical reactions due to reactive gas species are assisted by independently controlled energetic inert gas ions, has recently been developed. As shown in Fig. 4.22, this technique is carried out by an ion beam and a reactive gas pumped

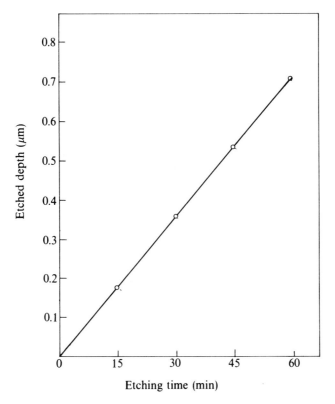

FIG. 4.21. Etched depth vs. etching time for $LiNbO_3$ [24]. CHF_3 gas pressure $= 8 \times 10^{-5}$ torr; ion energy $= 500$ eV; ion current density $= 0.4$ mA/cm^2

through a jet onto the workpiece in a work chamber. A number of qualitative and quantitative chemical models of the process have been proposed. For example, a model assuming implicitly a sputtering type of stock removal in Si and SiO_2, and a model assuming that surface damage exists in the work surface and promotes reaction of F atoms with SiO_2, have been proposed. Recently, Winter et al. [30] proposed a similar process of oxidation of Si and metals for ion-assisted etching of Si by F atoms.

Some special features of ion-beam-assisted chemical etching are summarized as follows:

(i) capability of a wide range of independent control of the chemical and ionic fluxes
(ii) controllability of the etched side wall profile
(iii) good etching selectivity and anisotropic etching
(iv) high etching rate.

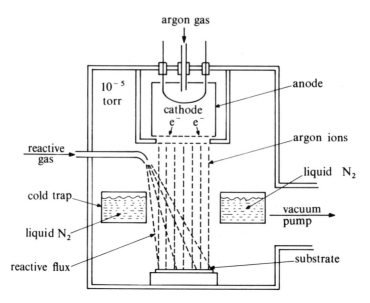

FIG. 4.22. Diagram of ion-beam-assisted etching (IBAE) system [30], consisting of an ion source, one or more reactive gas jets (only one shown), and a cold trap to pump the unused reactive gas and the reaction products. The gas jets are positioned ~ 3.5 cm from the sample

Characteristics such as the etching rate and etching selectivity have been reported for a wide variety of materials (Si, SiO_2, GaAs, refractory metals such as Mo and Ti, and refractory silicides such as $TiSi_2$, $MoSi_2$ and PtSi) with reactive gases (Cl_2, CCl_4 and XeF_2) and Ar as energetic ions.

The effects of ion energy, ion current density, reactive gas flow rate, and reactive gas pressure on these etching characteristics are now discussed.

The etching rate of Si vs. ion energy for incremental increases of Cl_2 gas under Ar^+ ion bombardment is shown in Fig. 4.23 [31]. It can be seen that for flow-rate increments of 3 cm^3 (volume at s.t.p.) of Cl_2, the etching rate of Si increases linearly. Rates were obtained up to five times that for the Ar ion beam alone. Higher etching rates could have been achieved by increasing the flow rate, but further chlorine addition is limited by the pumping speed of the vacuum system. This suggests that the etching rate of Si with Cl_2 under Ar ion bombardment is limited not by the desorption of silicon chlorides from the surface, as is the case in reactive ion etching, but rather by the availability of chlorine for reaction with Si.

4.2.4 Reactive ion etching (RIE)

Reactive ion etching is typically carried out in equipment of high-frequency (r.f.) type with opposed parallel target and base electrodes (see Fig. 1.24 in

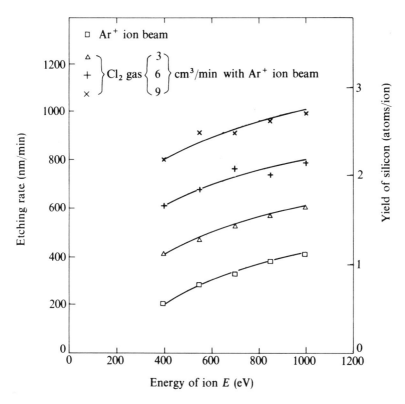

FIG. 4.23. Etching rate of Si vs. ion energy for incremental increases of Cl_2 gas under Ar^+ ion bombardment [31]. Ion dose = 0.04 C/cm^2; ion current density = 0.26 mA/cm^2

Chapter 1). The reactive gas is introduced at a pressure in the range 0.1 to 10 Pa and the r.f. power is increased between the target and the anode. The increased r.f. power at reduced pressure changes the reactive gas between the electrodes to plasma, ions of which impinge perpendicularly against the workpiece surface. The basic etching is caused by physical and chemical reactions between the incident reactive ions and the etched material, and chemical reaction between reactive radicals such as F^* (where the asterisk denotes the activated state) and the etched material. Reactive ion etching therefore has the advantages of physical ion-beam etching and of chemical plasma etching and will be applied widely to the etching processes in VLSI fabrication, as described in Chapter 1.

RIE is accomplished by introducing CCl_2 gas into h.f. or r.f. equipment with parallel plate electrodes. Etching of Si, SiO_2, glass, and Al at increased rates was first introduced by Hosokawa et al. [32]. Since Hosokawa's work, etching of Si and SiO_2 using CF_4 and Al using CCl_4 and BCl_3 gases has been developed. Recently, the RIE process has been widely used in the wafer process of compound semiconductor ICs such as GaAs and InP.

The main advantages and disadvantages of reactive ion etching are as follows. The advantages are:

(i) anisotropic etching
(ii) fairly good etching selectivity
(iii) high etching rate
(iv) capability of etching Al and SiO_2 at high rate
(v) side etching under a resist mask occurs under highly selective conditions or during long overetching.

Disadvantages are:

(i) carbonaceous films containing chlorine are formed during etching, which degrade the reproducibility of etching and corrode patterns after etching because of the high chlorine concentration, and are hard to remove even by oxygen plasma cleaning;
(ii) reactor chamber and rotary pump oil are heavily contaminated.

For micro-fabrication of VLSI, the important points are not only the etching rate of the substrate, but also the anisotropic etching and the etching selectivity between the substrate and films or resist. These characteristics depend on the combination of substrate and reactive gas species, r.f. power, pressure of reactive gas, and surface area of substrate. In the fabrication of solid-state electronic devices, silicon single crystal and polycrystal, SiO_2, Si_3N_4, and Al are the most important materials (see Table 4.2). Therefore, it is mainly the characteristics of these materials that have been investigated (see Table 4.3 [33]). Recently, the etching rates of compound semiconductors such as GaAs and InP have been reported [34].

The following sections describe the etching characteristics of Si, SiO_2, Al, and GaAs.

(1) Etching of Si

The etching of Si using CF_4 gas is explained in terms of the following reaction:

$$Si + 4F^* \rightarrow SiF_4 \uparrow. \tag{4.6}$$

That is, the etching of Si is accomplished by the chemical reaction between Si and F^*. In this case, F^* results in isotropic etching, and carbon formed by dissociation of CF_4 accumulates on the Si surface. As a result, the selective etching of Si on SiO_2 becomes impossible. The etching of Si is therefore performed by introducing oxygen, to remove the carbon film on the Si surface by the following reaction:

$$Si + O_2 + CF_n \rightarrow SiF_n \uparrow + CO, CO_2 \tag{4.7}$$

TABLE 4.2
Applications of reactive ion etching (RIE)

Device	Function	Substrate
1. Semiconductor devices		
IC	Conductor (pattern)	Poly-Si
Transistor	Electrode	Al, Al–Si
	Through-hole	PSG/SiO_2
	Insulator	Si_3N_4
2. IC mask	Metal mask	Cr/Cr oxide, W
3. Display devices		
LCD	Electrode	Al
PDP	Electrode	Cr/Cu/Cr
4. Miscellaneous		
Opto IC	Substrate	Glass, Al_2O_3
Bubble memory	Domain	
Thermal head	Heating element	Cr–Si, TaN
	Conductor	Cr–Cu, NiCr–Au

The etching rates of Si, SiO_2, and glass (7059), using CCl_2F_2 (halocarbon R–12), are shown in Fig. 4.24 [32]. The etching rate of Si is increased by a factor of two on addition of 10% oxygen, but at higher oxygen concentrations it decreases again. The initial increase is due to the removal of the carbon contamination resulting from the dissociation of the halocarbon ions. Excessive addition of O_2 results in reduction of the etching rate because of substrate oxidation.

The etching of Si using CF_4–O_2 causes undercutting of the pattern, so CF_4–H_2, CF_3Br, C_2F_6–CF_3Cl, and CF_4–Cl_2 have recently been tested [35, 36].

(2) Etching selectivity of Si and SiO_2 using CF_4 gas with I_2

The etching rate of various materials using CF_4 gas mixed with I_2 (20%) with a machining table made of carbon is shown in Fig. 4.25 [36]. In this case, the etching selectivity of Si vs. SiO_2 becomes greater than 6 to 8, with no undercutting. Moreover, Si, Mo, poly-Si, and Si_3N_4 have been selectively etched relative to SiO_2, similarly to conventional plasma etching, without any undercutting and with very high accuracy of pattern transfer. The etching of Si using CF_4 gas with I_2 is expressed by the following reaction:

$$Si + I + CF_n \rightarrow SiF_n \uparrow + CI \qquad (4.8)$$

TABLE 4.3
Etching gas and gas pressure for etching various materials [33]

Material	Etching gas	System	Operating pressure (Torr)[†]
Al	CCl_4, CCl_4+He, BCl_3 PCl_3, CCl_4+Cl_2 $SiCl_4$	Diode (A) Diode (C)	0.3 (CCl_4+He) 0.04 (CCl_4)
SiO_2	CF_4, CF_4+H_2, C_2F_6 C_3F_8, CHF_3	Diode (C)	0.02 (CHF_3)
Si	CF_4, CF_4+O_2, CCl_2F_2 $CSrF_3$, $SF_6+C_2H_4$ (CCl_4)	Diode (C) Cylindrical	0.1~0.5 (CF_4) 0.5~1.0 (CF_4+O_2)
Si_3N_4	CF_4, CF_4+O_2	Diode (C) Cylindrical	0.05~0.1 (CF_4) 0.1 (CF_4)
Mo	SF_6+CCl_4, $SF_6+C_2H_4$ CCl_4+O_2, CF_4+O_2 $CH_2Cl_2+O_2$	Diode (C) Diode (A)	0.4~0.5 0.6
Cr	CCl_4, CCl_4+air $CH_2Cl_2+O_2$	Diode (C) Cylindrical	0.003 0.3
Cr oxide	CCl_4, CCl_4+air $CH_2Cl_2+O_2$	Diode (C) Cylindrical	0.003 0.2
Au	CCl_4, CCl_2F_2	Diode (C)	0.03~0.001
SnO_2	$C_2H_5OH+air$, Ar CH_3OH	Diode (C) Cylindrical	0.008 ($C_2H_5OH+air$) 0.5 (CH_3OH)
In_2O_3	CH_3OH, CCl_4	Diode (C) Cylindrical	0.01 0.5
Nb	CF_4	Diode (C)	0.1
Ta	CF_4, CF_4O_2	Diode (C)	0.3
GaAs	CCl_2F_2, CF_4	Diode (C) Cylindrical	0.05 1~2.0

(A) = anode coupling, (c) = cathode coupling
[†] 1 Torr = 133 Pa

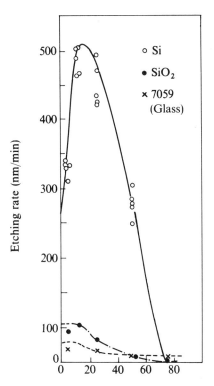

FIG. 4.24. Effect of oxygen on etching rate by CCl_2F_2 [32]. Etching time = 30 min; r.f. power density = 1.3 W/cm^2; gas pressure = 2.0×10^{-2} torr

As shown in Fig. 4.25, the etching rate increases with increasing r.f. power. This is due to the corresponding increases in plasma density (and hence reactive radical concentration) and ion energy.

The dependence of Si and SiO_2 etching rates, selectivity of Si vs. SiO_2, and V_{dc} (target voltage) on pressure for 40% Cl_2 in CF_4 are shown in Fig. 4.26. The etching rates and V_{dc} decrease with increasing gas pressure, while the etching selectivity of Si vs. SiO_2 increases with increasing gas pressure. Although pressure effects on these parameters are complex, the etching rate of Si is controlled by an ion-enhanced etching mechanism, since V_{dc} is inversely related to the chamber pressure.

(3) Etching of SiO_2

The etching of SiO_2 using CF_4 gas is explained by the following reaction:

$$SiO_2 + CF_n \rightarrow SiF_n \uparrow + CO, CO_2 \qquad (4.9)$$

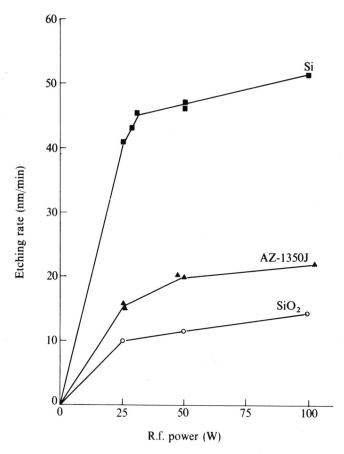

FIG. 4.25. Relation between RF power and etching rate of Si, SiO_2, and photoresist AZ-1350J, using CF_4 (15 cm^3/min) mixed with I_2 (20%), with a carbon etching table [36]. Pressure = 0.02 torr

The etching of SiO_2 requires a reaction assisted by ion bombardment. Therefore, to increase the etching selectivity of SiO_2 vs. Si, the etching rate of Si must be reduced by reducing the F:C ratio or the concentration of F and the etching rate of SiO_2 must be increased by the ion-beam-assisted effect. To reduce the concentration of F, H_2 gas has been added to CF_4, as shown in Fig. 4.27. The addition of H_2 reduces the etching rate of Si by scavenging fluorine radicals F* through the formation of HF. In this case, etching selectivities of SiO_2 vs. Si as high as 35 have been obtained, and anisotropic etching can also be achieved [37].

Trifluoromethane (CHF_3) is a selective plasma etchant for SiO_2 [38]. The discharge of reactive CHF_3 etches SiO_2 faster than Si, because a hydrogen-

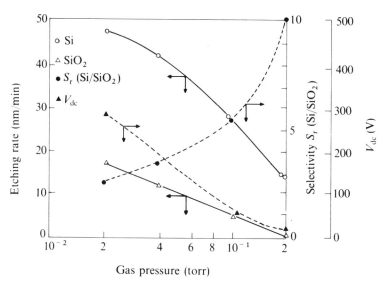

FIG. 4.26. Dependence of Si and SiO$_2$ etching rate, V_{dc} (target voltage), and selectivity S_r (etching rate ratio) on gas pressure at 40% Cl$_2$ in CF$_4$. R.f. power = 200 W; CF$_4$ + Cl$_2$ flow rate = 30 cm^3/min

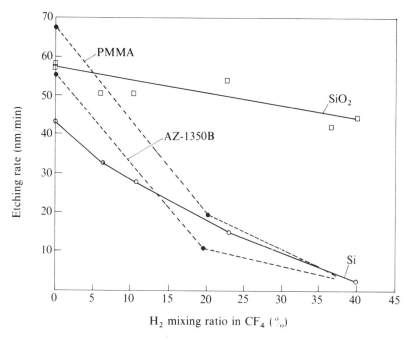

FIG. 4.27. Dependence of SiO$_2$, Si, PMMA, and AZ-1350B etching rates on H$_2$ mixing ratio in CF$_4$ [37]. R.f. power density = 0.26 W/cm^2; gas pressure = 4.7 Pa; gas flow rate = 28 cm^3/min

containing fluorocarbon plasma produces relatively little fluorine, the primary etchant for Si, but yields relatively large amounts of difluorocarbon radical, CF_2, a probable etchant for SiO_2. However, a shortcoming of hydrogen-containing etchants is that they produce deposited polymer films on the surface to be etched, and this results in a gradual decrease in the etching rate. For this reason, addition of NH_3 to CHF_3 has been studied. A plasma of CHF_3 with NH_3 does not deposit polymer film but maintains high selectivity of SiO_2 vs. Si and anisotropic etching.

(4) Etching of aluminium

For aluminium etching, it is possible to use a plasma of BCl_3 and CCl_4, because volatile aluminium chlorides can be made in both plasmas. However, there are some disadvantages in using CCl_4 gas for aluminium etching:

(i) Undercutting occurs under conditions of high etching selectivity.
(ii) The formation of carbonic films containing chloride on the surface to be etched results in the degradation of etching reproducibility or corrodes patterns.

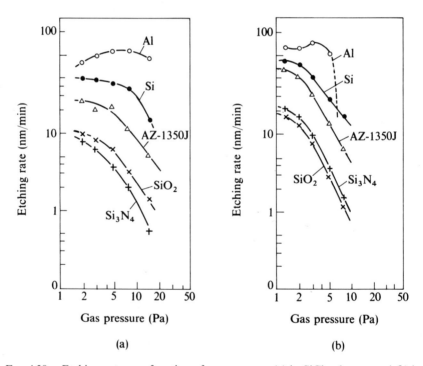

FIG. 4.28. Etching rate as a function of gas pressure, (a) in $SiCl_4$ plasma, and (b) in CCl_4 plasma [40]. R.f. power density = 0.16 W/cm^2

(iii) Etching chamber and rotary pump oil are heavily contaminated.

Recently, etching of aluminium using Cl_4–He [39] or $SiCl_4$ [40] gases has been developed. The gas pressure dependence of the etching rate of various materials is shown in Fig. 4.28 [40]. Aluminium is etched at high rate under a wider pressure range in $SiCl_4$ than in CCl_4, while the etching rates of other materials such as SiO_2, Si_3N_4, poly-Si, and photoresist decrease with increasing pressure of both gases, because the ion energy towards the workpiece surface decreases with increasing gas pressure. Therefore, high aluminium etching selectivity vs. other materials at a relatively high gas pressure can be obtained in all pressure ranges with $SiCl_4$ rather than with CCl_4. For example, the selectivities of Al vs. SiO_2 and Al vs. AZ-1350J are 35 and 10 respectively in $SiCl_4$ at 8 Pa.

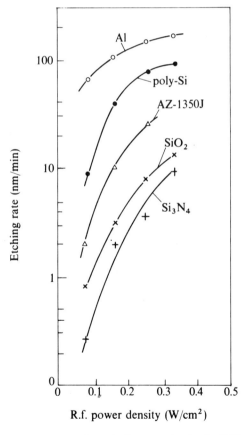

FIG. 4.29. Etching rate as a function of r.f. power density in $SiCl_4$ plasma [40]. Gas pressure = 8 Pa

The power density dependence of the etching rates of various materials in $SiCl_4$ at 8 Pa is shown in Fig. 4.29 [40]. The etching rate of aluminium shows less dependence on power density than do the rates of other materials. This indicates that the aluminium etching mechanism differs from that of the other materials. Aluminium is etched both by radicals and by the effect of ion bombardment, while the other materials are etched only by the effect of ion bombardment. This leads to higher etching selectivity in the lower power density region. As mentioned above, aluminium can be etched by using $SiCl_4$ and BCl_3 gases. However there are many problems such as induction time or dead time, loading effect and uniformity of etching.

4.2.5 Applications of ion-beam removal processes

Ion-beam removal processes can be applied in the fabrication of VLSI micropatterns, thinning of test pieces for transmission electron microscopy and of blanks in the manufacture of piezo-electric transducers, polishing or machining of hard and brittle materials such as ceramics, quartz, and diamond, and roughening of materials for biological implantation such as ceramic and calcium phosphate etc., as shown in Table 4.4. These applications are described below.

(1) Fabrication of solid-state electronic devices [41]

To obtain higher performance and increase the spatial density of gates in semiconductor integrated circuits, such as in VLSI, micro-fabrication technology for patterns of micrometre or submicrometre width are required. For example, in the fabrication of the lightly doped drain/source field-effect transistor (LDDFET), which has a channel width of 1.4 μm, reactive ion etching (RIE) is used to produce SiO_2 side wall spacers. The structure and fabrication steps of the LDDFET are shown in Fig. 4.30 [41]. The LDDFET device is first fabricated by a conventional process through the delineation of gate poly-Si by a chemical vapour deposition (CVD) SiO_2 etch mask. Reactive ion etching (RIE) with CCl_2F_2 reactive gas is used to etch both oxide and poly-Si so that, as shown in Fig. 4.30(b-2), vertical sides of the SiO_2/poly-Si gate stack are obtained. Next, 'n-regions' are formed by phosphorus ion implantation as shown in Fig. 4.30(b-3). A layer of CVD SiO_2 film of the desired thickness is then conformationally deposited as shown in Fig. 4.30(b-4), and using RIE with CF_4–H_2, the planar portion of the CVD oxide is removed. In these RIE steps, an end-point detection site is provided to minimize over-etching of the spacer oxide and gate stack. After the formation of the sidewall spacer, arsenic ion implantation is used to form the n^+ regions. During the n^+ ion implantation, the side wall effectively blocks arsenic ions

TABLE 4.4
Applications of ion-beam removal

Processing	Applications
Fabrication of integrated circuit & fine pattern	Solid-state electronic devices
	Magnetic bubble devices
	Surface acoustic wave filters
	Optical integrated circuits
	Optical gratings
	Zone plates
Thinning	Piezo-electric transducers
	Test pieces for transmission microscopy
Machining or polishing	Making of aspheric lenses
	Finishing of gauge blocks
	Forming of diamond tools
	Removal of strain caused by mechanical lapping
Surface roughening	Biological implants
	Solar cells
Drilling	Filters (mechanical)

from entering the two narrow n^- regions in the device as shown in Fig. 4.30(b-6). Thereafter, the device is fabricated by a conventional procedure up to the film steps.

This fabrication procedure facilitates excellent control and reproducibility of the n^- region of the LDDFET device. Moreover, the performance of the device is as much as 1.9 times that of conventional devices.

(2) Fabrication of magnetic bubble memories [42]

An X-ray lithographic process and ion-beam sputter etching are used to fabricate magnetic bubble memories of submicrometre diameter. The processes used to fabricate the X-ray mask and the magnetic bubble device using the mask are shown in Fig. 4.31. The positive master mask is patterned in PMMA (400 nm) photoresist by electron-beam lithography, and the gold X-ray absorber (300 nm) is electroplated. This positive master mask is patterned

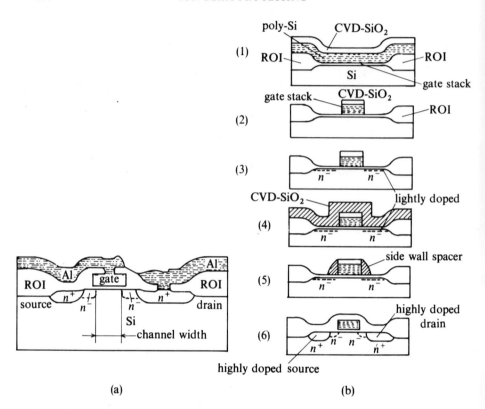

FIG. 4.30. (a) Structure of an LDDFET. (b) Fabrication steps: (1) formation of ROI, gate oxide and poly-Si–SiO$_2$ stack; (2) vertical RIE of gate stack; (3) n^- ion implantation; (4) deposition of SiO$_2$ for side wall space; (5) directional RIE to form side wall spacer; (6) n^+ ion implantation and dopant drive-in [41]. CVD = chemical vapour deposition

on a 2 μm thick polyimide substrate by X-ray lithography. The X-ray replication yields a positive pattern of Au, which is electroplated to produce a negative mask. The substrate for the negative mask is 4 μm polyimide and the device pattern is produced with an X-ray copy of this mask.

The fabrication steps for 1 μm bubble devices are shown in Fig. 4.31(b). Here, the resist is used as a template for electroplating permalloy on garnet. First a gold conductor is electroplated, and then the permalloy device is electroplated over the gold film. Finally, the Cr–Au base is etched using ion-beam sputter etching with Ar ions to leave isolated conductor elements in the circuit. By this procedure, magnetic bubble devices with 0.5 μm features can be fabricated.

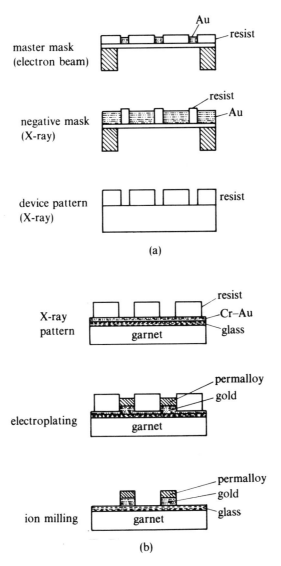

FIG. 4.31. (a) X-ray fabrication process. (b) SLM magnetic bubble process [42]

(3) Fabrication of surface acoustic wave filters [43]

A new fabrication process using ion-beam sputter etching has been developed for 0.5 μm finger period surface acoustic wave (SAW) filters, as shown in Fig. 4.32. A thick aluminium film 70 nm in thickness is deposited on a 128° YX-LiNbO$_3$ crystal. In step 1, the substrate (aluminium-plated LiNbO$_3$) is

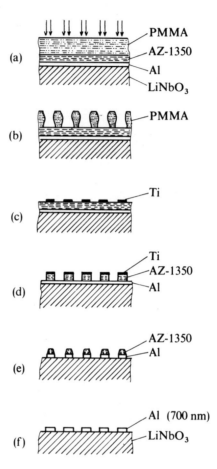

FIG. 4.32. Fabrication process sequence for submicrometre-wide electrode SAW filters: (a) electron-beam exposure; (b) PMMA development; (c) Ti lift-off; (d) AZ-1350 etching (oxygen ion beam); (e) Al etching (argon ion beam); (f) AZ-1350 removal (oxygen plasma)

coated with a photoresist film (AZ-1300) of 190 nm thickness, which is baked at 250°C for 1 h. Thereafter, a PMMA film of 450 nm thickness is coated, the pattern on the film is made by exposure to an electron beam, and it is developed in a 1:3 mixture of methyl isobutyl ketone (MIBK) and isopropanol (step 2). By the first process, PMMA patterns with undercut side walls are obtained. Next, a thick Ti film of 20 nm thickness is deposited on the sample surface by electron-beam evaporation. After the PMMA resist has been removed by acetone, Ti patterns are formed on the AZ-1300 film (step 3). These patterns are transferred into the AZ-1300 film (step 4), using ion-beam sputter etching with oxygen gas. Next, aluminium electrodes are delineated using ion-beam sputter etching with Ar ions through the Ti and AZ-1300 masks. Finally, the AZ-1300 mask is removed by oxygen plasma etching. By this process, low-loss SAW filters with electrodes of submicrometre dimension can be fabricated.

(4) Fabrication of thick zone plate [44]

Zone plates are used for diagonalizing X-ray and particle emission from laser-produced plasmas. To produce zone plates of virtually arbitrary thickness and high aspect ratio, reactive ion etching (RIE) and ion-beam etching (IBE)

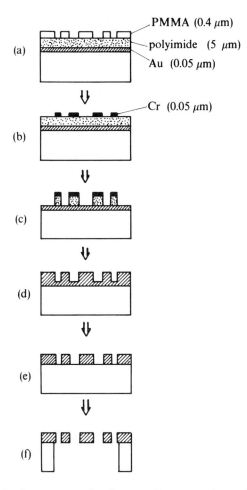

FIG. 4.33. Fabrication process of a free-standing zone plate using electron-beam lithography and reactive ion etching: (a) electron-beam lithography; (b) Cr lift-off; (c) O_2 reactive ion etching; (d) Au electro-plating, polyimide removal; (e) Ar ion-beam etching; (f) Si etching

have been used. The fabrication process for a free-standing zone plate using electron-beam lithography, RIE and IBE is shown in Fig. 4.33.

Silicon wafer substrate is successively coated with 10 nm thick nickel, 50 nm thick gold, 5 μm thick polyimide, and 0.4 μm thick PMMA. The pattern exposure of the PMMA is done with an electron beam of 80 nm dimeter and a digital scanning step of 60 nm. The chromium zone plate pattern is obtained by the lift-off process. The polyimide pattern of the plate is formed by RIE with oxygen gas, using the chromium pattern. Next, gold is electroplated in the etched polyimide grooves. After removal of the chromium by CCl_4 plasma etching, the polyimide is removed by oxygen plasma etching. Finally, the silicon substrate is back-etched using a 3:1 solution of HNO_3 and HF. An SEM photograph of a fabricated zone plate with gold about 3 μm thick is shown in Fig. 4.34.

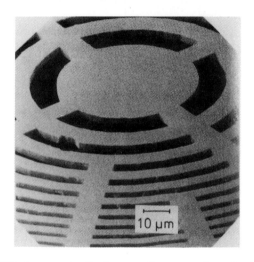

FIG. 4.34. SEM photograph of a free-standing zone plate of gold (3 μm thick)

(5) Fabrication of gratings

The micro-fabrication of $LiNbO_3$ for integrated optics can be achieved by Ar ion etching [45]. However, in this case, the accuracy of the waveguide profile is limited by mask shrinkage during ion etching, because the ratio of the etching rate of $LiNbO_3$ to that of mask material is very low. To obtain an accurate waveguide profile by increasing the etching rate of $LiNbO_3$ to AZ-1350 (photoresist), fabrication methods for SiO_2 blazed gratings [46] and for SiO_2 patterns with vertical side walls [47] using reactive ion-beam etching have been developed. These fabrication processes are shown in Fig. 4.35. An AZ-1300 photoresist of 0.13 μm thickness is coated on the $LiNbO_3$ substrate

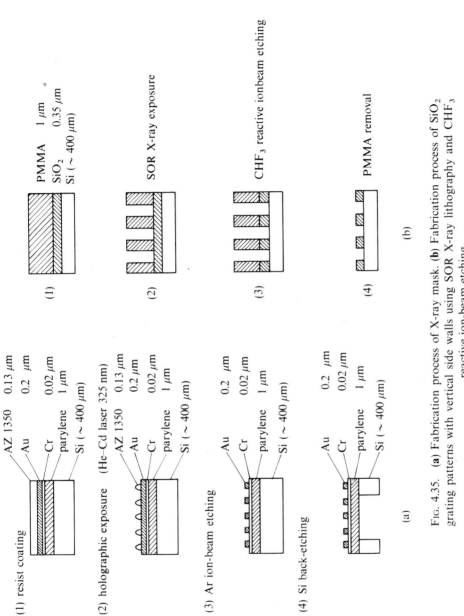

FIG. 4.35. (a) Fabrication process of X-ray mask. (b) Fabrication process of SiO$_2$ grating patterns with vertical side walls using SOR X-ray lithography and CHF$_3$ reactive ion-beam etching

and the film is exposed holographically using an Ne–Cd laser of 325 nm wavelength. After development, grating patterns are obtained. The blazed grating is then fabricated using CHF_3 reactive ion-beam etching.

(6) Fabrication of microlenses on an InP substrate [48]

A technique for fabricating microlenses on an InP substrate by ion-beam etching with Ar has been developed, and has been applied to fabrication of InGaAs–InP double heterostructure light-emitting diodes which have monolithic lenses on the light extraction surfaces. Such a diode and the sequence of the fabrication process for forming spherical lenses on an InP substrate are shown in Fig. 4.36. An AZ-1350 photoresist mask pattern with a diameter in the range 60–250 μm is delineated by conventional photolithographic technique. This is then baked at a temperature higher than the glass-transition point of the photoresist, so that spherical contours are generated by the effect of surface tension on the photoresist surface. Then the substrate is etched by an Ar ion beam. During the ion-beam etching, the spherical mask contour is transferred to the InP substrate surface. The lens shape is dependent on the initial mask thickness and the ratio of the etching rates of the substrate and the mask.

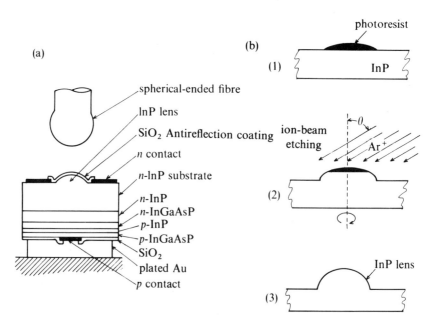

FIG. 4.36. (a) Structure of InP lens. (b) Fabrication process: (1) mask formation by patterning and baking a photoresist layer; (2) Ar ion-beam etching until mask completely removed; (3) completed InP lens

(7) Submicrometre pattern fabrication by focused ion beam [49]

Complex patterns of submicrometre feature size can be fabricated by focused ion-beam etching without conventional mask or resist processing. However, the throughput of this process is very low, because only a very small portion, with a dimension of ion beam size, is sputter-etched at a very low rate by physical sputtering. To increase the etching rate, therefore, ion-beam-assisted etching with a focused ion beam was developed. The application of focused ion-beam (FIB) technology for maskless ion implantation, etching, and deposition is described below.

(a) Si device fabrication by focused-beam direct ion implantation [50] A focused ion beam has been used for fabricating an Si submicrometre device. Two terminal resistors and p–n junctions are fabricated using a 60 keV Ga^+ focused ion beam of diameter as small as 0.2 μm. The FIB test structure and fabrication sequence of this test sample are shown in Fig. 4.37. After etching of alignment masks in the Si substrate, boron is implanted into 200 μm square pads. The sample is then annealed for 1 h at 1100°C in a nitrogen atmosphere. To identify the location of the boron-implanted regions, photoresist is applied and patterned into 100 μm squares over the boron pads. The Ga^+ ion beam is used to connect the boron pads, implanted 7° off the (100) axis to reduce channelling effects, by ten conducting lines, spaced 5 μm apart. After Ga^+ focused ion-beam implantation, the resist is removed and the sample is annealed at 800°C for 15 min in hydrogen. Aluminium is then evaporated and patterned on the 100 μm square contact pads.

(b) Fabrication of microbridge with focused ion beam [51] A fabrication process for a vertical-type microbridge using a focused ion beam has been developed, as shown in Fig. 4.38. Thin Si_3N_4 membrane is formed using anisotropic wet etching, and a small etch pit is formed using a 35 keV focused Ga^+ ion beam. After this process, a very thin layer of less than 20 nm is left for deposition of the first Nb layers. After the deposition of Nb film, ion-beam etching with Ar is performed *in situ* to etch away the remaining Si_3N_4 layer and to form a pinhole. Finally, a counter-electrode Nb film is evaporated to form the bridge.

(c) Mask repair using focused ion beam [52, 53] A scanning focused ion beam can be used to rework chromium-on-glass mask plates to repair clear defects such as pinholes, and opaque defects in the VLSI circuit pattern.

Repair of opaque defects can be performed by ion beam etching using a focused ion beam. Conversely, clear defects can be repaired with carbon film deposition using a focused ion beam. In this process, CVD gas is continually

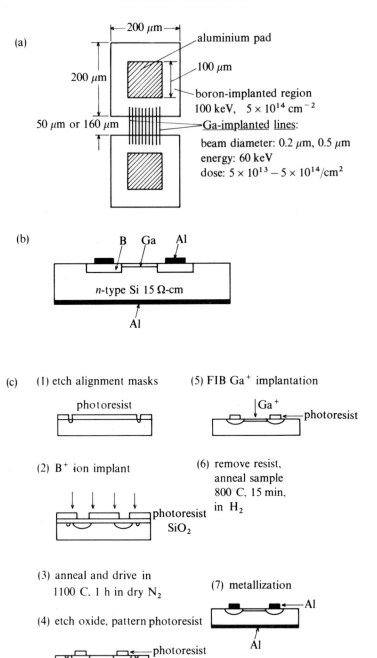

FIG. 4.37. (a), (b) Structure of FIB test sample. (c) Fabrication sequence [50]

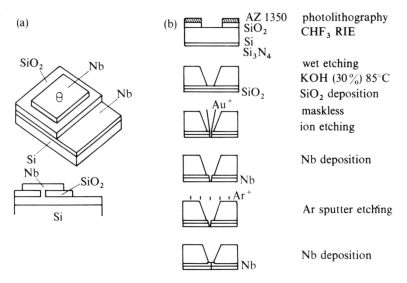

FIG. 4.38. (a) Vertical-type microbridge. (b) Fabrication process using a focused ion beam

jetted from a gas injector onto the mask surface defect area. The molecules are absorbed in the jetted area. A focused ion beam is scanned over the defect area, then, a selective deposition is formed by ion-beam-induced chemical reaction. An example of the mask repair is shown in *Fig. 4.39 [53]*.

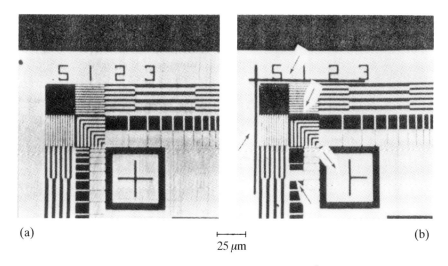

FIG. 4.39. Optical micrographs of mask before (a) and after (b) repair

(8) Finishing of gauge blocks of cemented tungsten carbide [54] and high carbon gauge steel [55, 56]

In contrast to the ion beam sputter-etching of hard and brittle materials like Si, glass, quartz and diamond, the ion beam sputter-etched surfaces of metals and sintered materials become rougher, especially at the stational ion incidence angle, even though the initial surface of metals and sintered materials have been mechanically polished to a minor-like finish. Because of the dependence of the etching rate on crystalline structure and axis of each crystal grain of such workpieces, adjacent crystal grains of the workpiece are eroded at different rates, which in turn develops overall relief or cone structure on the workpiece surface. In view of this difficulty, ion-beam sputter etching has been applied for the fine adjustment of dimensions of gauge blocks made of quenched steel and tungsten carbide steel. By rotation and swinging of the work table, the effective ion incidence angle to each crystal grain surface of the workpiece changes equally. A workpiece having the surface roughness of $R_{max} = 0.02$ μm can be produced by ion-beam sputter etching with the uniformly varying ion incidence angle system. By means of the same

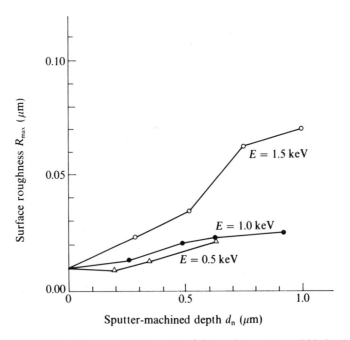

FIG. 4.40. Finishing of a gauge block made of sintered tungsten carbide by Ar^+ ion-beam etching, at different ion energies E and with uniformly varying incidence angle, θ, between 30° and 80°

technique, a workpiece of cemented tungsten carbide with a surface roughness $R_{max} < 0.03$ μm has been obtained, as shown in Fig. 4.40.

(9) Machining, sharpening and forming of diamond tools

Diamond tools such as styli, scribers, and indenters are usually polished mechanically, using fine diamond powder and a lap plate of soft material. In contrast, ion-beam sputter etching or machining is more effective in fine machining of hard and brittle materials such as diamond and ceramics.

The author has applied ion-beam sputter machining to the fabrication of diamond styli and diamond knives, and the sharpening of diamond point tools. Some examples are given below.

The profiling of a diamond workpiece (indenter) by ion-beam sputter machining with ion incidence vertical to the workpiece tip is illustrated in Fig. 4.41, which represents a two-dimensional section of a three-dimensional workpiece, symmetrical about the y axis of the x–y coordinates. The sputter machining rate $V_n(\alpha)$ of the straight line AB (inclined at an angle α to the horizontal) is greater than that of AO (the horizontal line). Therefore, if the surface lines of the profile are sputter-machined parallel to each surface by the ion beam, the profile of the workpiece becomes O'A'B'. The tip width of the

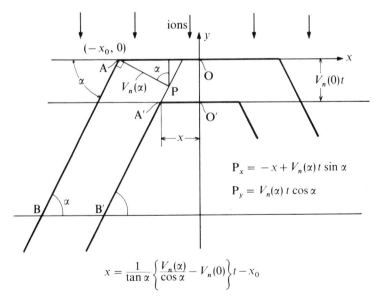

FIG. 4.41. Profiling of a diamond indenter by ion-beam sputter machining: calculation of length x of flat on the tip of the indenter

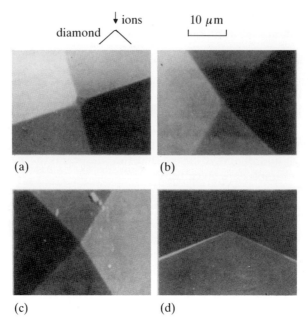

FIG. 4.42. Resharpening of an abraded Vickers micro-indenter by ion-beam etching [57]. $E = 1.0$ keV; $i = 0.39$ mA/cm^2. Sputter machining time: (a) 0 h; (b) 3 h; (c) 6 h; (d) 6 h viewed horizontally

indenter-like workpiece then decreases with machining time. Thus, resharpening of an abraded Vickers micro-indenter by ion-beam sputter machining or etching can be effected, as shown in Fig. 4.42 [57].

Scanning electron micrographs of a mechanically prefinished diamond stylus with a round tip of 10 μm radius, and of the ion-beam sputter-figured stylus, are shown in Fig. 4.43 [58]. Here the prefinished stylus has been subjected to ion sputter machining by an Ar ion beam of $E = 1.0$ keV to various tip radii. As shown in this figure, ion-beam sputter machining is readily applicable to forming the profile of a diamond stylus of 2 μm tip radius.

Similar photomicrographs of a mechanically prefinished diamond tool (microtome cutter) of 10 μm tip width and 1.0 mm tip length, and of one sharpened by ion-beam sputter machining with Ar ions, are shown in Fig. 4.44 [59]. As shown in this figure, ion-beam sputter machining can be used to form a diamond tool with a very sharp knife-edge shape.

(10) Surface roughening [60]

Texturing or roughening of material surfaces by ion-beam sputter etching will be useful and important in several different fields, such as controlled

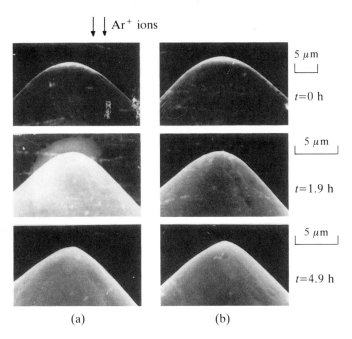

FIG. 4.43. SEM photographs of an ion sputter-machined diamond stylus observed in mutually perpendicular directions **a** and **b** at different machining times t [58]. $E = 1.0$ keV; $i = 0.50$ mA/cm^2; $\alpha = 44°$; $\gamma_0 = 10$ μm

thermonuclear fusion, solar energy conversion, and medicine. An application of surface roughening is in chemical cleaning of biological implants, to improve their performance and biocompatibility. The materials used for biological implants are classified into two groups: soft tissue implants (polyurethane and polyolefin) and hard tissue implants (Ti, Ti alloys, Co–Cr alloys, stainless steel, and alumina ceramic).

(11) Aspheric lens-making [61]

An aspheric lens-making system using ion sputter etching, in which the lens blank holder can swing and rotate independently in the working chamber, is shown diagrammatically in Fig. 1.23 (Chapter 1) and Fig. 4.45(a). A spherically polished lens, having a radius a little larger than the expected aspheric lens, is set initially and then subjected to ion-beam sputter etching with Ar ions of 20 keV. The height of the lens blank is adjusted so that the swing axis coincides with the centre of curvature of the lens surface. The centre of the swing axis is then offset to some extent from the centre of the ion beam, so that the ion beam can bombard the lens surface at a favourable angle. The centre of the angle of swing is aligned with the axis of rotation. To obtain the required aspheric surface, the rate of swing and the ion beam current density must be

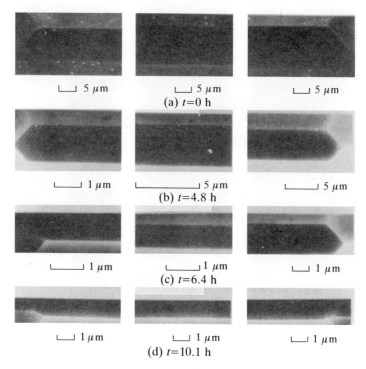

FIG. 4.44. Sharpening of knife-edge diamond tool (prefinished by mechanical lapping) by ion-beam etching with Ar^+: SEM photographs at different etching times t [59]. $l = 1.0$ mm; $E = 1.0$ keV; $i = 0.50$ mA/cm^2; $\alpha = 46°$

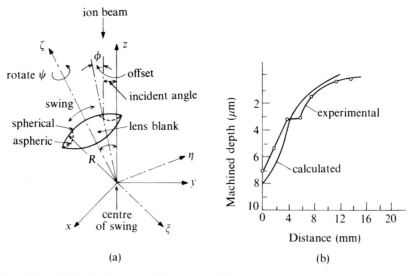

FIG. 4.45. (a) Aspheric lens-making system. (b) Measured and calculated machining profiles

controlled to remove a predetermined quantity from the lens blank. By this method, aspheric lenses of very smooth surface can be obtained. Some experimental data are shown in Fig. 4.45(b).

4.2.6 Problems of ion-beam removal processes

In the application of ion-beam processing to the machining of precise mechanical and optical parts and fabrication of solid-state electronic devices, the surface topography and radiation damage of the workpiece, and the accuracy of the obtained pattern profile of the substrate or the shape of the workpiece become important.

In the case of polycrystalline metals and alloys, and sintered powder materials, their surfaces are etched roughly by ion-beam etching, because the etching rates of individual crystal grains differ. Adjacent grains of the workpiece are eroded at different rates, which in turn develops an overall relief or cone structure on the surface. However, in the ion-beam etching of single-crystal and amorphous materials, if they are homogeneous and have clean surfaces, the surfaces are machined finely and uniformly by ion-beam etching. However, in both cases, a smooth surface can be obtained by combining an oblique ion incidence of about 60° with rotation of the workpiece.

When ion-beam etching processes are applied to the fabrication of the ultra-fine patterns of VLSI, an altered or damaged layer of the substrate results from the ion bombardment. This process of induced radiation damage can become a serious problem. To minimize the thickness of the damaged layer, the substrate must be sputter-machined with ions of energy lower than 200 eV at an ion incidence angle greater than 60°.

The pattern profile of the substrate is affected by factors such as undercutting, redeposition of sputtered materials, faceting (Fig. 4.46) [62] of the top edges of the pattern and trenching (Fig. 4.47) [52] at the bottom of the pattern. When the ultra-fine pattern is delineated by ion-beam etching processes, these factors must particularly be taken into consideration. Faceting becomes a serious problem in the fabrication of a diamond knife by ion-beam etching.

The curve of angular dependence of the machining or etching rate always has a maximum. This maximum in the V_n–θ (depth machining rate vs. incidence angle) curve results in the formation of facets with an apex angle corresponding to the ion incidence angle at which the etching rate is greatest. The effect of redeposition becomes very important, particularly along the edges of steep pattern features, and is sometimes observed along the edges of a photoresist pattern, where ridges of material remain when the resist pattern is removed. The last important factor, predominantly encountered near surface features with steep sides, is enhanced erosion due to ion reflection at the side wall. Combined bombardment by the primary ion beam and the reflected ion

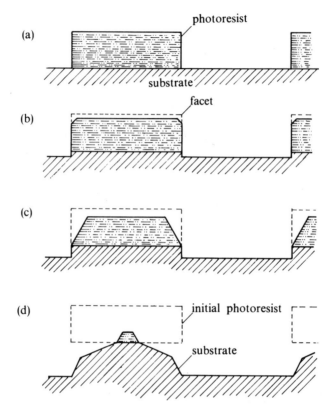

FIG. 4.46. Facet formation in photoresist during ion-beam etching [91]. (a) Photoresist cross-section before etching. (b), (c) Facet formation. (d) Further etching causes new facet formation in substrate

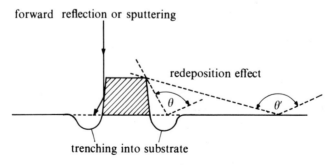

FIG. 4.47. Trench formation at the side of patterns [92]

beam at the inclined side wall enhances the etching rate and results in groove formation at the bottom of cones, pyramids, and ridges.

4.2.7 Simulation of changes in surface topography and profile of the workpiece

The surface topography or profile of the workpiece can be developed by ion-beam etching using increased machining times. The simulation of changes in surface topography or profile during ion-beam etching has been investigated by various authors along with the delineation of ultra-fine patterns of VLSI and the development of the surface topography of the workpiece.

First, along with the development of the surface topography, the motion of the intersection of two planes during ion-beam etching, and the formation of the facet at the side wall of a surface step were investigated [63]. Then the evolution of surface shape of amorphous materials by ion-beam etching was analysed [64]. Subsequently this method was extended to the computer simulation of the development of the sinusoidal surface wave during ion-beam etching [65]. More recently a three-dimensional treatment of the evolution of surface topography, based on the method of characteristic trajectory, was extended to the appearance and development of edges and facets on amorphous and crystalline surfaces undergoing ion-beam etching [66].

4.3 Ion-beam deposition

4.3.1 Ion sputter deposition

The ion sputter deposition method has the disadvantages of complexity of the equipment and low deposition rate compared with ordinary vacuum evaporation methods. However, this method has the advantage of being able to deposit materials of high melting point and low vapour pressure.

In ion sputter deposition, surface atoms of a target subjected to ion bombardment are usually considered as the source of deposition material from which films on a substrate in a glow discharge domain are grown. Figure 4.48 shows typical d.c. ion sputtering equipment. The target which is to be deposited is connected to a high negative voltage, so that when glow discharge is started, the target is bomarded by positive ions from the plasma. Then atoms are sputtered from the surface of the target and deposited on the substrate mounted opposite it. The substrate holder is usually earthed and sometimes heated or cooled. In general, axial and transverse magnetic fields are applied to increase the ionization efficiency by elongating the path of the ionizing electrons.

A target made of insulating material is not etched by d.c. sputtering, because ions are reflected by the positive potential formed on the target

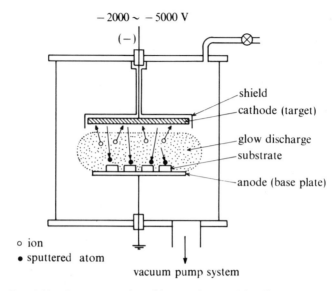

Fig. 4.48. Ion sputter deposition equipment (d.c. discharge type)

surface by the accumulation of positive ions. For this reason, radio-frequency equipment, as shown in Fig. 1.24 (Chapter 1), has been developed. However, this equipment has the disadvantage of low deposition rate, one-tenth of that obtained by vacuum evaporation. High-performance sputter coating equipment using magnetron oscillation has been developed for applications ranging from microelectronics (see Table 4.5) to architectural glass plates.

Magnetron sputtering equipment can have a variety of geometric configurations, as shown in Fig. 4.49 [67]. Magnetron sources can provide high current density due to the $E \times B$ field configuration (electric field perpendicular to magnetic field), which forms an intense plasma region near the target (cathode). Therefore, a high sputter deposition rate is obtainable even at relatively low pressure. The magnetic plasma confinement greatly reduces the temperature of the substrate, compared with conventional sputtering deposition equipment, and permits a heat-sensitive substrate such as plastic and high-polymer sheet to be coated at high deposition rates.

To increase the uniformity of the film thickness and step coverage of a pattern, the substrate must be supported in a planetary or other movable support system and rotated. For this purpose, planar and S-gun magnetrons [68] have been developed. In these types of equipment, permanent magnets are mounted behind the target, so the thickness of the target is limited. However, there is no restriction on substrate location and movement.

In magnetron-type equipment, the sources are operated at high current and low voltage, and relatively low pressure, less than 0.1 Pa, so the sputtered

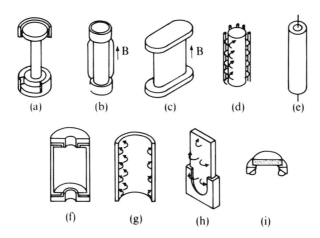

FIG. 4.49. Schematic illustrations of various types of magnetron sputtering source [72]. (a) Cylindrical post magnetron with electrostatic end confinement. (b) Cylindrical post magnetron with magnetic end confinement. (c) Rectangular post magnetron. (d) Ring discharge post magnetron. (e) Spiral discharge post magnetron. (f) Cylindrical hollow magnetron with electrostatic end confinement. (g) Ring discharge hollow magnetron. (h) Planar magnetron. (i) S-gun magnetron

atoms from the target are deposited on the substrate surface without colliding with residual gas atoms. Hence the deposition rate of film at the substrate is proportional to the ion current density at the target and depends secondarily on the substrate location. Examples of the deposition rates of various materials are given in Fig. 4.50 [69], which shows that the deposition rate increases linearly with increasing d.c. or r.f. power. The dependence of deposition rate on working gas pressure is shown in Fig. 4.51 [69]; the deposition rate has a maximum value at pressures ranging from 0.2 to 0.5 Pa. Some examples of the deposition rates of various materials are summarized in Table 4.6 [69].

Using a compound target, a thin film of the compound can be obtained by the sputter deposition method. However, in this case the composition of the thin compound film obtained differs slightly from that of the target compound. For the deposition of oxides, therefore, oxygen is introduced into the working Ar gas to obtain the desired film composition. By this method, the deposition and quality of the film can be controlled by adjusting the reactive gas composition in the working Ar gas. In this case, the formation of the surface compound on the target surface, which often results in a significant reduction in deposition rate, becomes a problem. Figure 4.52 shows a planar configuration which is used to deposit compounds at a reasonably high rate in the absence of the formation of compounds on the target surface.

TABLE 4.5
Applications of sputter deposition in electronics

Device	Application		Substrate
Integrated circuit	Beam lead	(Pattern)	Pt–Ti–Au, Al
	Conductor	(Electrode)	Al, Cr–Cu
	Circuit	(Resistor)	Al, Cu, Au, Al–Si
		(Resistor)	NiCr, Cr, Ta, TaN, Ta–Si
		(Capacitor)	Ta_2O_5, SiO_2, Al_2O_3, TiO_2
Electronic parts	Circuit (Functional)	(Resistor)	NiCr, Thermet
	Magnetic head	(Passivation)	SiO_2
	Video disc/audio disc	(Electrode)	Al
		(Passivation)	SiO_2
	Magnetic disc	(Memory)	γFe_2O_3, Co–Cr/Ni–Fe
	Thermal head	(Heater)	$TaSiO_2$, Cr–Si, Ta_2N
		(Abrasion resistance)	Ta_2O_5, SiC
	Bubble memory	(Memory)	Ni–Fe
		(Conductor)	Al–Cu
		(Insulation)	SiO_2

	Josephson junction	(Memory)	NiFe
		(Conductor)	Cu
	Thin film	(Memory)	Ni–Fe
	Magnetic memory	(Conductor)	Cu
Chromium mask	Hard mask		Cr/Cr_2O_3
ITO film	Transparent conductor		In_2O_3, SnO_2
	Insulation		Al_2O_3
	Passivation		SiO_2, MgO
Electro-luminescent device	Phosphor		ZnS, ZnSe, CdS
	Conductor		Al
	Transparent conductor		In_2O_3, SnO_2
	Passivation		Y_2O_2, SiO_2, Si_3N_4
Plasma display panel	Conductor		Cr–Cu–Cr
	Insulation		Al_2O_3, SiO_2
	Passivation		MgO
Printed circuit board			Cu

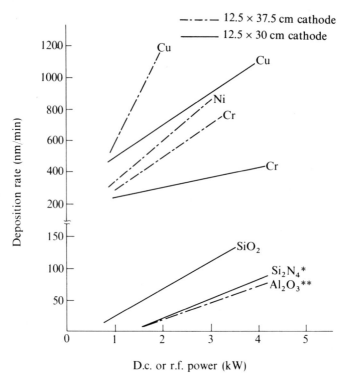

FIG. 4.50. Sputter deposition rates of various cathode materials (metal targets by d.c. mode, others by r.f. mode) [69]. *Reactive sputtering of Si in Ar and N_2. **Reactive sputtering of Al in Ar and O_2. Pressure $= 5 \times 10^{-3}$ torr (6.5×10^{-1} Pa)

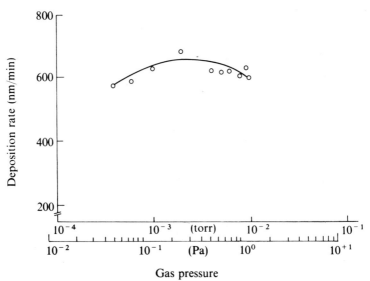

FIG. 4.51. Sputter deposition rate vs. gas pressure for a Cu target [69]. D.c. power $= 1$ kW; target–substrate distance $= 60$ mm

TABLE 4.6
Typical sputtering yields and deposition rates [69]

Element	Sputtering yield (atoms/ion)	Average deposition rate (nm/min)
Ag	3.4	2650
Al	1.2	760
Au	2.8	2200
C	0.2 (Kr)	160
Co	1.4	—
Cr	1.3	1000
Cu	2.3	1800
Fe	1.3	—
Ge	1.2	770
Mo	0.9	700
Nb	0.65	500
Ni	1.5	—
Os	0.95	740
Pd	2.4	1870
Pt	1.6	1260
Re	0.9	700
Rh	1.5	1170
Si	0.5	400
Ta	0.6	470
Ti	0.6	470
U	1.0	800
W	0.6	470
Zr	0.75	600

The coating properties such as structure and stress in the sputter-deposited thin film are strongly affected by the incident angle of depositing atoms to the substrate, bombardment by energetic working gas atoms and energetic ions, and the substrate temperature relative to the melting point (T_m) of the coating film. The influence of substrate temperature and Ar working gas pressure on the structure of metal coated film which is deposited by cylindrical magnetron sputtering is shown in Fig. 4.53 [70]. Zone 1 consists of tapered crystals which are poorly bonded. Zone T is a transition structure consisting of a dense array of poorly defined fibrous grains. Zone 2 is characterized by evolutionary growth due to adatom diffusion and consists of columnar grains separated by distinct dense intercrystalline boundaries. Zone 3 consists of equiaxed grains and results from bulk diffusion processes such as recrystallization. A smooth

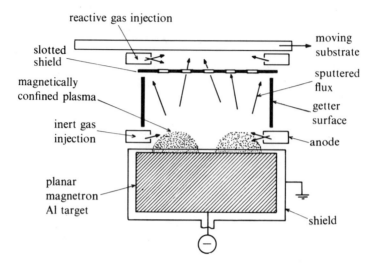

FIG. 4.52. Schematic drawing of a planar magnetron configuration used to deposit compounds at reasonably high rates without poisoning. The slotted shield is used to permit a reactive gas concentration gradient to be maintained between the substrate and the cathode [67]

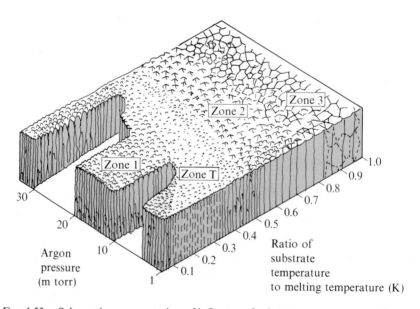

FIG. 4.53. Schematic representation of influence of substrate temperature and argon pressure in the working chamber on structure of metal coatings deposited by sputtering using a cylindrical magnetron source [70]

substrate surface, the arrival of the coating atoms in a direction perpendicular to the substrate surface, and bombardment of energetic ions and working gas atoms promote the formation of the dense zone T coating structure. Zone T coating is generally in a state of compression. On the other hand, oblique incidence due to geometry effects at low pressure, or gas scattering at high pressure, promote more open structure and tensile stress.

4.3.2 Ion-beam sputter deposition

Ion-beam sputter etching equipment, such as the ion-beam shower type and focused ion-beam type, can be used for thin film deposition. That is, atoms sputtered from the target by ion-beam bombardment are deposited on the substrate surface. This deposition method allows isolation of the substrate from the plasma from which ions are extracted.

In ion-beam sputter deposition, the working chamber can be maintained at low pressure, of the order of 0.01 Pa, because the ion beams are extracted from the ion source. The substrate or thin film is therefore kept at low temperature and pure films of the highest quality are obtainable.

The equipment described in section 4.2 is used for ion-beam sputter deposition. Sometimes, dual ion-beam equipment, as shown in Fig. 4.54 [71], is used in a single chamber. One gun is used for ion-beam sputter deposition, while the other is used to bombard simultaneously the substrate surface; the material deposited on the substrate can then be modified by interaction with the ion beam from the other gun. This method is used to produce crystalline films on the substrate and diamond-like carbon films.

By introducing the reactive gas into the working chamber or using reactive ions as shown in Fig. 4.55 [72], it is possible to deposit compounds such as oxides and nitrides.

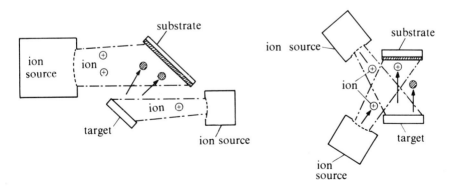

FIG. 4.54. Dual beam configurations for sputter deposition [71]

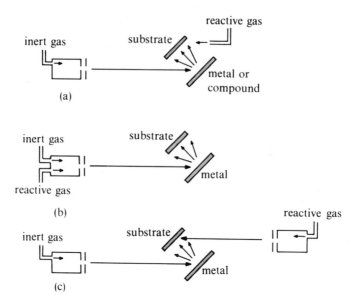

FIG. 4.55. Reactive secondary ion-beam deposition methods [72]. (a) Inert gas ion beam, reactive gas added to substrate region. (b) Reactive and/or inert gas ion beam. (c) Dual ion-beam method using inert gas ion beam to sputter metal target and reactive gas ion beam to bombard growing film

Ion-beam sputter deposition can be used to form various kinds of films, but this method is used mainly for investigating the mechanism of growth of sputtered thin films and for obtaining high-quality thin films.

4.3.3 Ion plating

Ion plating, first reported by Mattox [73] in 1963, is physical vapour deposition, in that the deposited layer on the substrate is bombarded by ions during vacuum coating due to èvaporation, as shown in Fig. 4.56. In general, vapours of materials to be deposited on the substrate are generated by a resistance-heated or electron-beam-heated crucible source, and diffuse from the vapour source to the substrate through an inert gas such as Ar. During the process, the inert gas in the working chamber is maintained at a pressure in the range 1 to 10 Pa. Therefore, when a high voltage is applied between the substrate (cathode) and another earthed electrode (anode) within the vacuum system, a glow discharge between the substrate and an anode is produced. As a result of this glow discharge, the substrate is surrounded by a space charge region, across which a potential difference of about 100 V occurs or is applied. Ar ions and ionized atoms of vapour materials then form the discharge and

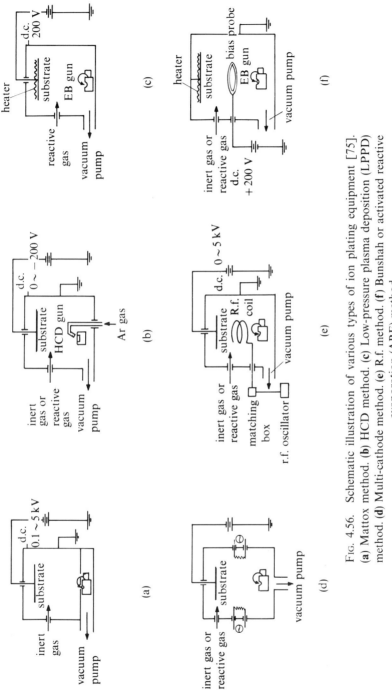

FIG. 4.56. Schematic illustration of various types of ion plating equipment [75]. (a) Mattox method. (b) HCD method. (c) Low-pressure plasma deposition (LPPD) method. (d) Multi-cathode method. (e) R.f. method. (f) Bunshah or activated reactive evaporation (ARE) method

TABLE 4.7
Comparison of various ion plating methods [74]

Plating method	Ionization method	Pressure (Pa)	Evaporation source	Ambient gas	Acceleration of ions	Temperature of substrate
Mattox	Glow discharge by d.c. high voltage	0.5~1	RHF (EB)	Inert	600~5 kV Dependent on ionization voltage	High
HCD	Bombardment by low voltage and high current electron beam	0.01~0.1	HCD	Inert Reactive	0~200 V Independent of ionization voltage	Low
LPPD	Low-voltage glow discharge between EB gun and substrate	0.01~0.1	EB	Reactive	None	Low (heater)

Multi-cathode	Bombardment by electrons emitted from hot cathodes	$10^{-3} \sim 0.1$	RHF EB	(Reactive)	$0 \sim 5$ kV Independent of ionization voltage	Low (heater)
RF	Glow discharge by r.f. coil (high voltage)	$0.01 \sim 0.1$	RHF EB	Inert Reactive	$0 \sim 5$ kV Independent of ionization voltage	Low
ARE	Low-voltage glow discharge between EB gun & bias probe	$0.01 \sim 0.1$	EB	Reactive C_2H_2, CH_4 N_2, O_2	None	Low (heater)

RHF = resistance-heated filament, EB = electron-beam gun, HCD = hollow-cathode discharge, LPPD = low-pressure plasma deposition, RF = radio frequency, ARE = activated reactive evoporation bombard the substrate, while the vapour material is simultaneously deposited on the substrate.

This method has the advantage of good adhesion or consolidation between the deposited thin film and the substrate, and also of good coverage on surfaces out of the line of sight of the vapour source (high throwing power). The good consolidation results from the bombardment of the substrate by high-energy atoms and ions, and the high throwing power from scattering by residual gas atoms of the atoms to be deposited. However, deposited films are generally required to be as free as possible from impurities introduced by the deposition environment, and to adhere strongly to the substrate. To achieve the first condition it is desirable to deposit the film rapidly in a clean vacuum system at a low pressure.

Since the introduction of Mattox's technique, there have been a number of developments and improvements. In particular, increased ionization of the depositing material has been studied. These improved techniques are shown in Table 4.7 [74] and Fig. 4.56 [75].

When a resistance-heated filament or boat is used as the heat source of the evaporator, only materials of relatively low melting point can be evaporated, and films of only a few micrometres thickness can be obtained. Lately, therefore, an electron-beam gun has been used as the heat source, by means of which, all metals of high melting point, such as tungsten, titanium, and molybdenum, can be deposited at rates of over 10 μm/min. However, in this case, a differential pumping system must be used, because the ion plating requires a pressure below 1 Pa, whereas an electron-beam gun with a hot filament requires a pressure less than 0.05 Pa.

Refractory materials such as TiN, TiC and WC cannot be vaporized. To deposit refractory metal nitrides and carbides, reactive ion-plating techniques such as activated reactive ion plating (ARIP) have been developed. For example, TiC is synthesized by the reaction of Ti metal vapour and C_2H_2 gas molecules on the substrate surface. As shown in Table 4.8 [76], various compounds can be synthesized by reactive ion plating.

TABLE 4.8
Compounds synthesized by ARE ion plating process [76]

Oxides	α-Al_2O_3, γ-Al_2O_3, Y_2O_3, Ti oxides
Carbides	TiC, ZrC, NbC, Ta_2C, TaC, VC, W_2C, HfC, VC–TiC, TiC–Ni
Nitrides	Ti_2N, TiN
Sulphides	$Cu_xMo_6S_8$, Cu_xS
Other	Nb_3Ge

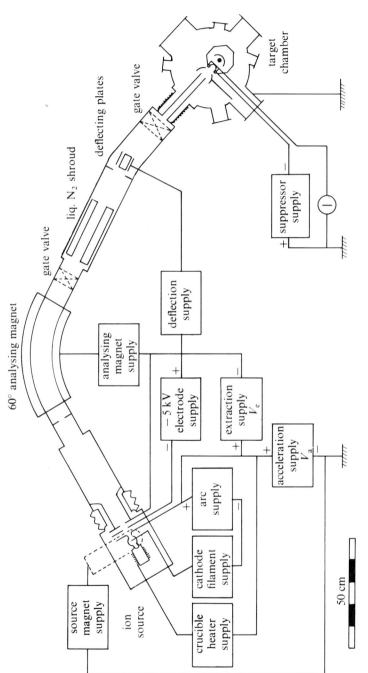

FIG. 4.57. Ion-beam direct deposition system

4.3.4 Ion-beam direct deposition [77]

Ion-beam direct deposition equipment is shown in Fig. 4.57. Ions generated in the ion source are selected by mass/energy ratio by magnetic means, and then decelerated by a static lens system. In this way, pure and defined ion beams which have low energy, in the range 10 to 100 eV, can be deposited on the substrate. The energy of the ion beam is below or too much above the threshold for point defect creation and sputtering.

The ion-beam direct deposition method is used mainly as a tool for fundamental research into the elementary mechanism of ion-activated thin film deposition such as ion plating. However, it can also be applied to the epitaxial synthesis of diamond, and to deposit silver thin film on a substrate. The pure silver thin film deposited on a mild steel substrate has a low friction coefficient and good loading characteristics, as shown in Fig. 4.58 [78].

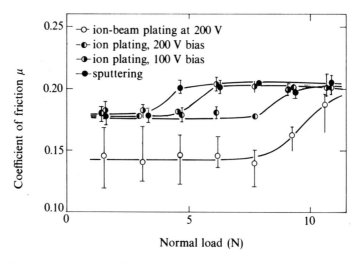

FIG. 4.58. Coefficient of friction of vacuum-deposited silver films (sliding speed = 0.5 mm/s) [78]

4.4 Ion-beam surface treatment

4.4.1 Ion nitriding [79, 80]

Ion nitriding (plasma nitriding) of alloy steels containing aluminium, chromium, and vanadium, as a surface hardening treatment, is carried out using a glow discharge at a low pressure (10 to 1000 Pa) of nitrogen–hydrogen gas mixture or NH_3 gas at a temperature between 500 and 540°C. The temperature of the workpiece is raised by bombardment with ions formed in the glow discharge plasma and is below the eutectoid temperature of 590°C in the binary iron–nitrogen alloy system. Nitrides of aluminium, chromium, and vanadium are also effective in surface hardening. Therefore, no shape distortion resulting from a phase change to martensite can occur. The autothermic nature of the process and absence of shape distortion are major benefits of ion nitriding. Moreover, this process has a number of other distinct advantages, compared with gas nitriding and other heat treatments, which may be summarized as follows:

(i) considerable reductions in processing time and in consumption of energy and processing gas are achieved;
(ii) scale on the workpiece surface can be removed by ion bombardment;
(iii) the quality and thickness of the compound or nitride layer can be controlled, so the treated workpiece has good mechanical properties;
(iv) areas which do not need to be treated can be mechanically masked.

However, the process has the following disadvantages:

(i) the initial cost of the equipment is high compared with that of gas nitriding equipment;
(ii) thorough degreasing of the workpiece before ion nitriding is required.

The process is now widely used in the automobile industry, mould and die industry, polymer processing industry, etc. Recently, focused ion-beam nitriding has been investigated.

4.4.2 Ion-beam implantation [81, 82]

Ion-beam implantation is a method in which a broad or finely focused energetic ion beam, in the range 10 to 200 keV, is used to introduce dopants into the surface layer of solid materials. Ion-beam implantation has been used extensively in the manufacture of semiconductor devices, in which group V (P, As) and group IV (B) ions are now routinely used for the n^+ and p^+ doping of Si devices.

The principal advantages of ion implantation over thermal diffusion technology are as follows:

(i) small lateral spreading of dopants
(ii) very high dose reproducibility and controllability
(iii) low-temperature processing
(iv) automation capability.

In the past, only low-current ($< 10 \mu A$), and medium-current (0.1–0.5 mA) ion implanters have been used. However, owing to the advance of VLSI circuits and devices, high-dose and high-throughput ion implanter systems

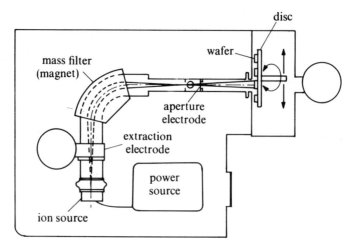

FIG. 4.59. Schematic view of a low-energy, high-current, batch process ion implanter

have become necessary. Recently, a high-current implanter (10 mA), shown in Fig. 4.59, has been developed for this purpose. As shown in the drawing, ion implanters usually have an ion source, an ion extraction system, and a mass separator. After ion-beam analysis, only the desired ion species is accelerated to its final energy, focused, and electrostatically scanned uniformly over a target. For detailed explanation of this process, the reader should refer to a specialist work [83].

Recently, application of ion implantation to the modification of the chemical or mechanical properties of materials which are not used in electrical devices has been examined.

4.4.3 Ion-beam mixing [84, 85]

Ion-beam mixing is the process in which ion beams are used in conjunction with conventional thin film deposition techniques. That is, a thin film is first deposited on the substrate by conventional deposition techniques. After deposition, the substrate is bombarded by a high-energy ion beam (above 200 keV) through the surface thin film, so that when ions are implanted in the substrate through the thin film, the atoms of the film, bombarded by incident ions, diffuse into the interface between the thin film and the substrate: the ion-beam mixing process.

An extension of this idea involves simultaneous ion bombardment and thin film deposition. This combination allows the production of a thicker alloyed region than does either direct ion implantation or ion-beam mixing.

At present, the ion-beam mixing process is used for the formation of thermal-equilibrium and metastable silicide phases, the formation of metastable alloys, epitaxial growth of Pd_2Si film, etc.

4.5 Other applications of ion-beam processing

4.5.1 Ion-beam lithography or pattern exposure [86]

Ion-beam lithography can be carried out by the use of a broad ion beam as the exposing iradiation through a mask placed in proximity to a resist-coated wafer, or by the use of a finely focused ion beam for direct writing in a manner similar to the scanning electron beam.

For broad ion-beam lithography, three types of mask have been considered: a stencil mask, an amorphous thin film membrane, and a somewhat thicker single-crystal film mask membrane oriented to allow the ions to channel; in these cases scattering of ions in the mask membrane is the ultimate limiting feature of broad ion-beam lithography.

For fine focused ion-beam lithography, or microfine ion-beam lithography with a beam less than 1 µm in diameter, it can be expected that fine patterns with a size comparable with or smaller than the ion beam can be delineated even in a resist layer on thicker wafer, because impinging ions suffer only slight scattering compared with electron beams, and because of the small range of secondary electrons. Moreover, ion-beam lithography provides a very high-speed exposure technique, owing to the large energy deposition rate of ion beams in resist materials. By both ion-beam lithography methods, 40 nm lines have been replicated in PMMA, using a gold mask or finely focused ion-beam.

The use of a focused ion-beam is expected to be one of the most useful techniques in micro-fabrication in the next decade. In particular, ion-beam lithography will be a leading pattern delineation technique for pattern sizes near to and less than 0.1 µm. Microfine focused ion beams will be used for maskless ion implantation (doping), maskless ion etching, maskless ion-beam-assisted chemical etching, maskless deposition of materials onto a substrate, and so on.

4.5.2 Ion-beam chemical analysis [87, 88]

A number of surface analytical techniques have been developed which are based on the detection of particles such as backscattered ions, secondary ions, neutral ions, photons, X-rays, etc, emitted from a solid as a result of its interaction with bombarding or incident ions.

When solids are bombarded with an ion beam of energy ranging from 1 to 50 keV, positively or negatively charged secondary ions, and neutral atoms which are emitted from the solid due to sputtering, can be analysed using a mass spectrometer: these techniques are referred to as secondary ion mass spectrometry (SIMS) and sputtered neutrals mass spectrometry (SNMS), respectively. When the bombarding ion beam has a diameter of several micrometres and is rastered over the sample, the former technique is frequently referred to as ion microprobe mass analysis (IMMA). This technique has a very high sensitivity and spatial resolution power for elements with high ionization efficiencies, owing to the small sampling depth.

A method based on the detection and energy analysis of particles backscattered from a solid as a result of its interaction with an He^+ ion beam of high energy (above 1 MeV) is referred to as high-energy ion scattering (HEIS) or Rutherford backscattering (RBS). Using this technique, it is possible to obtain in-depth information about impurities and disorder of the lattice in the surface layer of the solid.

Other surface analytical techniques based on ion bombardment are summarized in Table 4.9.

TABLE 4.9
Analytical technique based on ion (atom) bombardment

	Excitation conditions	Detected species	Acronym
Photon detection	1–10 keV, Ar^+, O_2^+, O^-	Visible light	BLE
	2 MeV He^+	X-rays	PIXE
	Gas discharge*	Visible light	GDOES*
Electron detection	5–10 keV Ar^+	Energy of Auger electrons	IIAES
	1–10 keV, Ar^+, O_2^+, O^-, Cs^+	Sputtered secondary ions	SIMS
	1–10 keV atoms		FAB
	Gas discharge*		GDMS*
Ion detection	1–10 keV ion beams or gas discharge	Post-ionized neutrals $[(m/e)°]^+$	SNMS
	1–5 keV He^+ Ne^+	Reflected projectile energy	LEIS
	100 keV He^+		MEIS ISS
	2 MeV He^+		HEIS (RBS)

* Not a beam technique

References

[1] Pivin, J. C., *J. Mater. Sci.* **18** (1983), 1267.
[2] Matsuo, S., *Jpn. J. Appl. Phys.* **21** (1982), L4.
[3] Brown, W. L., *Proc. Int. Ion Engg. Congress* (1983), 1738A.
[4] Komuro, M. et al., *J. Electrochem. Soc.* **126** (1979), 483.
[5] Chinn, J. D. et al., *J. Vac. Sci. Technol.* **B1** (1983), 1028.
[6] Thompson, M. W., *Phil. Mag.* **18** (1968), 377.
[7] Sigmund, P., *Phys. Rev.*, **187** (1969), 383.
[8] Brand, W. et al., *Nucl. Inst. Meth.* **47** (1967), 201.
[9] Pivin, J. C., *J. Mater. Sci.* **18** (1983), 1267.
[10] MacDonald, R. J., *Adv. Phys.* **19** (1970), 475.
[11] Almén, O. et al., *Nucl. Instr. Methods* **11** (1961), 279.
[12] Hosaka, S. et al., **18** (1975), 384 (Japanese).
[13] Lee, R. E., *J. Vac. Sci. Technol.* **16** (1979), 164.
[14] Carter, G. et al., *Ion bombardment of solids* (Heinemann Educational Books, London, 1968).
[15] Behrisch, R., editor, *Sputtering by particle bombardment I* (Springer-Verlag, Berlin, 1981).
[16] Cantagel, M. et al., *J. Mater. Sci.* **8** (1973), 1711.
[17] Vossen, J. L., *J. Phys. E: Sci. Instrum.* **12** (1979), 159.
[18] Bunshah, R. et al., *Deposition technologies for films and coatings: developments and applications* (Noyes Publications, Park Ridge, 1982), 188.
[19] Matsuo, S. et al., *Jpn. J. Appl. Phys.* **21** (1982), L4.
[20] Heath, B. A., *J. Electrochem. Soc.* **129** (1982), 397.
[21] Powell, R. A., *Jpn. J. Appl. Phys.* **21** (1982), L170.
[22] Kaki, K. et al., *11th Symp. on Ion implantation and sub-micron etching* (Inst. for Phys. & Chem. Res., 1982), 31.
[23] Matsui, S. et al., *Jpn. J. Appl. Phys.* **20** (1981), 1735.
[24] ———, *Jpn. J. Appl. Phys.* **19** (1980), L463.
[25] Okano, H. et al., *ibid.* **20** (1981), 2429.
[26] Horiike, Y. et al., *ibid.* **18** (1979), 2309.
[27] Mayer, T. M. et al., *J. Electrochem. Soc.* **129** (1982), 585.
[28] Okano, H. et al., *Jpn. J. Appl. Phys.* **21** (1982), 696.
[29] Matsui, S. et al., *ibid.* **20** (1981), L38.
[30] Winter H. F. et al., *J. Vac. Sci. Technol.* **B1** (1983), 463.
[31] Chinn, J. D. et al., *ibid.* **B3** (1985), 410.
[32] Hosokawa, N. et al., (1974). *Jpn. J. Appl. Phys.* **Suppl. 2, Pt. 1** (1974), 435.
[33] Matsumoto, T. (1985). *Sputtering systems, dry etching systems and its applications* (Catalog of the ULVAC Corporation, Tokyo, 1985).
[34] Kurogi, Y., *Proc. of the 5th Sym. on Ion Sources and Ion-Assisted Technology (ISIAT), Kyoto* **15** (1981).
[35] Matsuo, S., *Appl. Phys. Lett.* **36** (1980), 768.
[36] Shibagaki, M. et al., *Jpn. J. Appl. Phys.* **19** (1980), 1579.
[37] Ephrath, L. M., *J. Electrochem. Soc.* **126** (1979), 1419.
[38] Chinn, J. D., *J. Vac. Sci. Technol.* **19** (1981), 1418.

[39] Nishimura, H. et al., *Shinku* **25** (1982), 624 (Japanese).
[40] Sato, M. et al., *J. Vac. Sci. Technol.* **20** (1982), 186.
[41] Tsang, P. J. et al., *IEEE J. Solid-State Circuits* **SC-17** (1982), 220.
[42] Hofer, D. et al., *J. Vac. Sci. Technol.* **16** (1979), 1968.
[43] Itoh, M. et al., ibid. **21** (1982), 21.
[44] Aritome, H. et al., ibid. **B3** (1985), 265.
[45] Kaminow, I. P. et al., *Appl. Phys. Lett.* **24** (1974), 622.
[46] Matsui, S. et al., *Jpn. J. Appl. Phys.* **19** (1980), L463.
[47] Matsui, S. et al., ibid. **20** (1981), 1735.
[48] Wada, O., *J. Electrochem. Soc.* **131** (1984), 2373.
[49] Muray, J. J., *Proc. Int. Ion Engg. Congress, Kyoto* (1983), 1545.
[50] Hamadeh, H. et al., *J. Vac. Sci. Technol.* **B3** (1985), 91.
[51] Namba, S., *Proc. Int. Ion Engg. Congress, Kyoto* (1983), 1533.
[52] Heard, P. J. et al., *J. Vac. Sci. Technol.* **B3** (1985), 87.
[53] Kaito, T. et al., *Proc. 9th Symp. on ISIAT '85 Tokyo* (1985), 207.
[54] Taniguchi, N. et al., *Annals of the CIRP* **30** (1981), 499.
[55] Taniguchi, N. et al., ibid. **24** (1975), 125.
[56] Miyamoto, I. et al., *Proc. of the 3rd ICPE Kyoto* (1977), 145.
[57] Miyamoto, I. et al., *Prec. Engg.* **4** (1982), 191.
[58] —— ibid. **5** (1983), 61.
[59] —— *Bull. Japan Soc. of Prec. Engg.* **17** (1983), 195.
[60] Weigand, A. J. et al., *J. Vac. Sci. Technol.* **14** (1977), 326.
[61] Taniguchi, N. et al., *Annals of the CIRP* **23** (1974), 47.
[62] Lee, R. E., *J. Vac. Sci. Technol.* **16** (1979), 164.
[63] Stewart, A. D. G., *J. Mater. Sci.* **4** (1969), 56.
[64] Carter, G. et al., ibid. **6** (1971), 115.
[65] Catana, C. et al., ibid. **7** (1972), 467.
[66] Smith, R. et al., ibid. **42** (1980), 235.
[67] Thorton, J. A., *Proc. 6th Symp. on ISIAT '82* (Tokyo, 1982), 363.
[68] Vossen, J. L. et al., editors, *Thin film process* (Academic Press, 1978), 497.
[69] Catalogues of the ULVAC Corporation (Tokyo).
[70] Thorton, J. A., *Deposition technologies for films and coatings: depositions and applications* (Noyes Publications, Park Ridge, 1982), 214.
[71] Reader, P. D., *Proc. 6th Symp. on ISIAT* (Tokyo, 1982), 29.
[72] Vossen, J. L. et al., editor, *Thin film process* (Academic Press, 1978), 196.
[73] Mattox, D. M., *J. Appl. Phys.* **34** (1963), 2493.
[74] Ohtsuka, H., *Kinzoku Hyomen Syori Gijyutsu* **35** (1984), 25 (Japanese).
[75] Katayama, S., *Kikaino Kenkyu* **33** (1981), 1146 (Japanese).
[76] Bunshah, R. F., *Proc. Int. Workshop by Professional Group on Ion-Based Techniques for Film Formation* (Kyoto, 1981), 311.
[77] Tsukizoe, T. et al., *J. Appl. Phys.* **48** (1977), 4770.
[78] Ohmae, N., *J. Vac. Sci. Technol.* **13** (1976), 82.
[79] Denton, R. A., *J. Vac. Sci. Technol.* **8** (1971), 1.
[80] Jones, C. K. et al., *Metal Progress* **85** (1964), 94.
[81] Cater, G. et al., *Ion implantation of semiconductors* (Edward Arnold, London, 1976).
[82] Turner, N. L., *Ionics* **8** (Tokyo, 1982), 120.

[83] Dearnaley, G. et al., *Ion implantation* (North-Holland, Amsterdam, 1973).
[84] Nicolet, Marc-A. et al., *Mat. Res. Soc. Symp. Proc.* **27** (1984), 3.
[85] Furukawa, S. et al., *Proc. Int'l Ion Engg. Congress, Kyoto* (1983), 1817.
[86] Bartelt, J. L., *J. Vac. Sci. Technol.* **19** (1981), 1166.
[87] Wittmack, K., *Vacuum* **34** (1984), 119.
[88] Werner, H. W. et al., ibid. **34** (1984), 83.
[89] Almen, O. et al., *Nucl. Instr. Methods* **11** (1961), 257.
[90] Mayer, T. M. et al., *J. Vac. Sci. Technol.* **21** (1982), 757.
[91] Smith, H. I., *Proc. IEEE* **62** (1974), 1361.
[92] Melliar-Smith, C. M., *J. Vac. Sci. Technol.* **13** (1976), 1008.

APPENDIX A: TEMPERATURE ANALYSIS OF THERMAL ENERGY PROCESSING BY ELECTRON BEAM AND LASER BEAM

To clarify the fundamental mechanism and characteristics of thermal processing by energy beams such as the laser beam and the electron beam, the temperature distribution in the solid workpiece due to the input energy must first be determined. However, the practical thermal processing of materials involves phase changes; and moreover, the thermal properties of materials vary with working temperature. Accordingly, it is difficult to determine the temperature distribution in the workpiece from ordinary heat conduction theory.

In this section, the heating characteristics of the workpiece for a temperature rise to the melting point are discussed, where the thermal properties are assumed to be constant. Furthermore, considerations of latent heat accompanied by phase change will be discussed in special cases.

NOTATION

Roman symbols

a	radius of circular beam or major radius of an elliptic section of melted zone (m)
b	half-width or minor radius of an elliptic section of melted zone (m)
c	specific heat capacity (J/kg K)
C_σ	coefficient of surface tension with respect to temperature (N/m K)
d	circular beam diameter, $=2a$ (m)
D	hole depth (m)
H	penetration depth (m)
L_m	latent heat of fusion (J/kg)
m	non-dimensional radial distance r/w, or distance from surface z/w
q	surface input heat flux per unit time per unit area (W/m^2)
Q	total input surface heat flux per unit time (W)
r	radial distance from incident-beam centre on workpiece surface (m)
R_p	electron range (m)
t	time (s)
t_c	characteristic response time, given by $\pi a^2/\kappa$ (s)
t_r	pulse repetition time (s)
T	temperature rise (K)
T_m	melting point (K)
T_0	initial temperature or reference temperature at infinite distance (K)

T_s	saturation temperature at beam center on workpiece surface (K)
U	speed of beam, material removal, or welding (m/s)
w	standard-deviation radius in Gaussian power density distribution (m)
x, y, z	coordinates (m)
z	depth (m)

Greek symbols

β	non-dimensional time, $\kappa t/w^2$
κ	thermal diffusivity, $\lambda/\rho c$ (m²/s)
λ	thermal conductivity (W/m K)
μ	viscosity (Pa s)
ν	kinematic viscosity, μ/ρ (m²/s)
ρ	mass density (kg/m³)
τ	pulse duration (s)
ϕ	heat generation in target material per unit time per unit volume, or volumetric power density (W/m³)

A.1 Basic equation of thermal energy processing based on heat conduction theory

If a heat source ϕ exists inside the workpiece material, the heat conduction equation is given by

$$\rho c \frac{\partial T}{\partial t} = \frac{\partial}{\partial x}\left(\lambda \frac{\partial T}{\partial x}\right) + \frac{\partial}{\partial y}\left(\lambda \frac{\partial T}{\partial y}\right) + \frac{\partial}{\partial z}\left(\lambda \frac{\partial T}{\partial z}\right) + \phi \quad (A.1)$$

When thermal properties ρ, c, and λ vary with temperature, eqn (A.1) is non-linear.

A.2 Temperature rise in a semi-infinite solid due to a surface heat source

Projected laser beams are absorbed by a very thin surface layer of solid material except in transparent materials, whereas electron beams penetrate some distance into a solid surface layer. The penetration depth of 100 keV electrons is of the order of 10^1 μm; for example, 60 μm in aluminium. The two cases cannot therefore be dealt with similarly. For $R_p \ll a$ or w, electron-beam heating can be treated as heating due to a surface heat source.

A.2.1 Axially symmetrical power density of input energy beam

The basic, boundary, and initial equations in a semi-infinite solid for constant thermal properties are given by

$$\frac{\partial T}{\partial t} = \kappa\left(\frac{\partial^2 T}{\partial r^2} + \frac{1}{r}\frac{\partial T}{\partial r} + \frac{\partial^2 T}{\partial z^2}\right) \quad (A.2)$$

$$-\lambda \frac{\partial T}{\partial z} = q(r, t) \quad \text{at } z = 0 \quad (A.3)$$

$$T = 0 \quad \text{as } r, z \to \infty \quad (A.4)$$

$$T = 0 \quad \text{when } t = 0 \quad (A.5)$$

If $q(r, t)$ is uniform and constant inside a circular region, i.e. $q = Q/\pi a^2$, where a is the beam radius, the temperature rise $T(r, z, t)$ is given [1] by

$$T(r, z, t) = \frac{aq}{2\lambda} \int_0^\infty J_0(\xi r) J_1(\xi a) \left\{ e^{-\xi z} \operatorname{erfc}\left[\frac{z}{2\sqrt{\kappa t}} - \xi\sqrt{\kappa t}\right] \right.$$
$$\left. - e^{\xi z} \operatorname{erfc}\left[\frac{z}{2\sqrt{\kappa t}} + \xi\sqrt{\kappa t}\right] \right\} \frac{d\xi}{\xi} \quad \text{(A.6)}$$

where J_0 and J_1 are Bessel functions of zero and first orders, and erfc is defined as

$$\operatorname{erfc} x = \frac{2}{\sqrt{\pi}} \int_x^\infty e^{-\xi^2} d\xi$$

The calculated results for eqn. (A.6) have already been discussed partly in section 1.4.3.

A.2.2 Gaussian distribution of power density of input energy beam

If $q(r, t)$ is independent of t but obeys the Gaussian distribution, i.e. $q(r) = Q\exp[-(r/w)^2]/\pi w^2$, the temperature rise T at the workpiece surface ($z = 0$) is expressed, using non-dimensional temperature $T\lambda w/Q$, as follows [2]:

$$\frac{T\lambda w}{Q} = \frac{1}{\pi^{3/2}} \int_{\tan^{-1}(1/2\sqrt{\beta})}^{\pi/2} e^{-m^2 \sin^2 \eta} d\eta \quad \text{(A.7)}$$

where $m = r/w$. At the beam centre ($m = 0$),

$$\frac{T\lambda w}{Q} = \frac{1}{\pi^{3/2}} \tan^{-1} 2\sqrt{\beta} \quad \text{(A.8)}$$

The temperature rise T along the normal to the surface at the centre, i.e. the z axis, is given by

$$\frac{T\lambda w}{Q} = \frac{1}{\pi^{3/2}} \int_{\tan^{-1}(1/2\sqrt{\beta})}^{\pi/2} e^{-m^2 \tan^2 \eta} d\eta \quad \text{(A.9)}$$

where $m = z/w$. The calculated results are given in ref. [2]. Some examples are shown in Figs. A.1 and A.2.

The temperature distribution in the workpiece reaches a saturation state as $t \to \infty$ ($\beta \to \infty$). For a beam of Gaussian power density distribution, the saturation temperature T_s (K) at $r = z = 0$ is given by

$$T_s = \frac{Q}{2\sqrt{\pi}\lambda w} \quad \text{(A.10)}$$

For a uniform circular beam, T_s is given by

$$T_s = \frac{Q}{\pi \lambda a} \quad \text{(A.11)}$$

which is shown in Chapter 1.

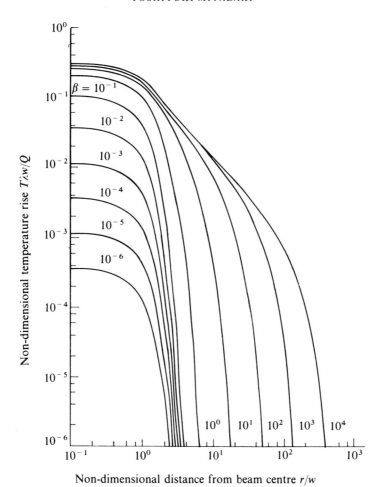

Fig. A.1. Temperature distribution on the surface of a semi-infinite workpiece due to a stationary Gaussian beam [2]

A.2.3 *Energy beam conditions required for machining a localized portion of a workpiece*

These conditions have already been discussed numerically in Chapter 1; therefore, an analytical expression for the temperature rise on the workpiece surface is given here for evaluating the necessary conditions qualitatively. From eqn. (A.7), for $\beta = \kappa t/w^2 \ll 1$,

$$T = \frac{2}{\lambda} \frac{Q}{\pi w^2} e^{-(r/w)^2} \sqrt{(\kappa t/\pi)} \qquad (A.12)$$

The quantity $Q \exp[-(r/w)^2]/\pi w^2$ is the power density of the Gaussian distribution; therefore, the temperature rise profile is similar to that of the input power density for a

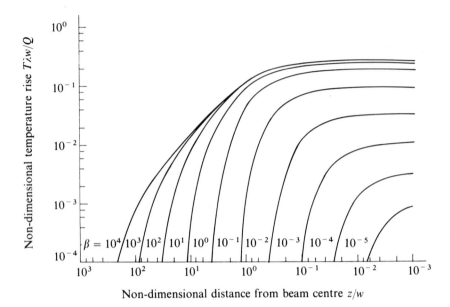

FIG. A.2. Temperature distribution along the normal to the surface of a semi-infinite workpiece due to a stationary Gaussian beam [2]

given workpiece and heating time. When the heating time becomes long, the heat diffuses around the projected area, and the temperature rise occurs more widely, as shown in Chapter 1. To concentrate the temperature rise within a restricted portion or the area where the beam power is supplied, a beam of higher power density, corresponding to a higher value of $Q \exp[-(r/w)^2]/\pi w^2$, should be used, because the temperature rise T in eqn. (A.12) must reach the melting point or vaporization point within a short heating time t.

A.3 Temperature rise in a plate due to a surface heat source

In section A.2, the workpiece was assumed to be semi-infinite. In the thermal energy-beam machining of plates, the workpiece should be considered to be a plate which extends infinitely, but whose thickness is of the order of the beam diameter. In such a case, the considerations in Chapter 1 and the temperature rise equation in section A.2 cannot be applied to the temperature analysis. A simple method is to modify the equation in the semi-infinite solid for this case, using the method of images.

For a plate whose thickness is D and whose rear surface is insulated thermally, i.e.

$$\frac{\partial T}{\partial z} = 0 \quad \text{at} \quad z = D \tag{A.13}$$

the temperature rise T' is given as follows [3]:

$$T' = \sum_{n=-\infty}^{\infty} T(r, 2nD + z, t) \qquad (A.14)$$

where T is the temperature rise in a semi-infinite solid. When the plate thickness is 10 times the beam radius or more, the temperature rise can be roughly evaluated for practical use from the equation for a semi-infinite solid.

A.4 Consideration of the process based on electron penetration

For a fine, high-energy electron beam, the heat source inside the workpiece due to electron penetration must be considered. The power density ϕ due to the incident electron beam as shown in eqn. (A.1) depends on the properties of the material, because the electron penetration range R_p varies with the material properties. For simplicity, a circular heat source is considered, whose radius and thickness correspond to the beam radius of 0.1 mm and the electron penetration range R_p, respectively. The values of R_p for 100 keV electrons in copper, aluminium, and gold are about 20, 60, and 10 μm respectively [4–7]; therefore, for an electron beam of 1 kW, values of ϕ are 1.6×10^3, 5.3×10^2, and 3.2×10^3 W/m^3 for copper, aluminium, and gold respectively. Temperature distributions involving electron penetration are given in refs. [8–13].

Electron penetration range is an important parameter, especially in the processing of thin films or foils. The incident electron energy is absorbed in the workpiece along the path of the electrons; consequently, a thin film is heated according to the law of two-dimensional heat conduction, and a thick foil is heated according to three-dimensional conduction.

Isothermal planes due to heat conduction in aluminium and copper for a temperature rise of 1000 K are shown in Fig. A.3 under the conditions shown [13]. In a workpiece considerably thinner than the electron penetration range, the radial distances at which the isothermal planes expand do not depend on the thickness. The isothermal plane expands more widely in copper than in aluminium. Though the thermal conductivity of copper is higher than that of aluminium, more rapid temperature rise occurs in copper. This fact can be inferred only from the difference in the effective volumetric power densities for the two materials due to the electron penetration. The effective volumetric power density for copper is 2.6 times that for aluminium.

A.5 Effects of a pulsed beam on the heated zone

As discussed in sections 1.4.3 and A.2, a rapid temperature rise in a localized portion can be obtained by a beam of high power density with a very short heating time. An example of the effects of a repeated pulsed beam on the heated zone is shown in Fig. A.4 [14]. A semi-infinite block of zirconium is heated for 1 s by a strip-like beam 3 mm in width, with 50 Hz pulses of duration 0.01, 1, and 10 ms, in rectangular wave form. In calculating the temperature distributions, the power density per unit area on the workpiece surface is determined so that the temperature rise at the centre line of the

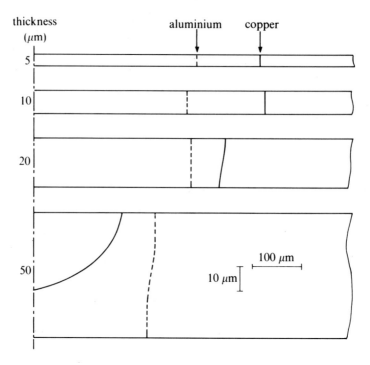

FIG. A.3. Isothermal planes (temperature rise = 1000 K) for various thicknesses after heating for 30 μs by an electron beam of 100 kV and 35 mA, with $w = 0.38$ mm [13]

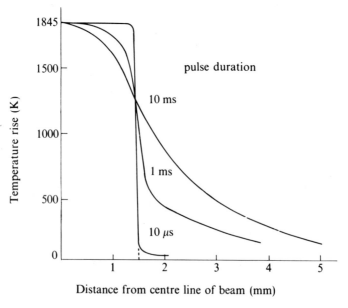

FIG. A.4. Effect of pulsed beam on concentration of temperature rise [14]

strip on the workpiece surface reaches the melting point (2118 K) at the end of a heating time of 1 s. The temperature distributions in Fig. A.4 are along the normal to the centre line on the surface. A temperature rise within a more restricted region is obtained by a beam of shorter pulse duration.

Optimum pulse conditions for concentrating the temperature rise in a restricted region have been obtained under a circular surface heat source with uniformly distributed thermal power density on a semi-infinite solid. For the steady state due to a pulsed input beam of rectangular wave form, the mean temperature rise $\bar{T}(r, t)$ on a hemisphere of radius r in the workpiece is given [15], using nondimensional form $\bar{T}\lambda a/Q$, by

$$\frac{\bar{T}\lambda a}{Q} = \frac{1}{\pi}\left[\frac{\tau}{t_r}\left(1-\frac{r}{2a}\right)+2\sum_{n=1}^{\infty}\frac{\sin\left(\frac{n\tau}{t_r}\pi\right)}{n\pi}\mathbb{R}\left\{\frac{1-e^{-\alpha_n r}-e^{-\alpha_n a}\sinh(\alpha_n r)}{(\alpha_n a)(\alpha_n r)}e^{jn\omega t}\right\}\right] \quad (A.15)$$

for $r \leqslant a$,

$$\frac{\bar{T}\lambda a}{Q} = \frac{1}{\pi}\left[\frac{\tau}{t_r}\frac{a}{2r}+2\sum_{n=1}^{\infty}\frac{\sin\left(\frac{n\tau}{t_r}\pi\right)}{n\pi}\mathbb{R}\left\{\frac{\cosh(\alpha_n a)-1}{(\alpha_n a)(\alpha_n r)}e^{-\alpha_n r}e^{jn\omega t}\right\}\right] \quad (A.16)$$

for $r > a$, and

$$\frac{\bar{T}\lambda a}{Q} = \frac{1}{\pi}\left[\frac{\tau}{t_r}+2\sum_{n=1}^{\infty}\frac{\sin\left(\frac{n\tau}{t_r}\pi\right)}{n\pi}\mathbb{R}\left\{\frac{1-e^{-\alpha_n a}}{\alpha_n a}e^{jn\omega t}\right\}\right] \quad (A.17)$$

for $r=0$, where $\mathbb{R}$ represents the real part of the complex number $j=\sqrt{-1}$, $\omega=2\pi/t_r$, and $\alpha_n = (1+j)\sqrt{(n\omega/\kappa)/2}$.

Calculated results for the sharpness of temperature distribution at the end of each pulse, i.e. the ratio of temperature rise at a point of radius r to that at $r=0$, are shown in Fig. A.5 for $\tau/t_r = 0.01$ and 0.1. It is clear that the sharpness of the temperature distribution depends not only on the pulse duty factor (τ/t_r), but also on the ratio of the repetition cycle time to the characteristic response time ratio (t_r/t_c). The thermal diffusivity of the material plays an important role in determining t_c, since $t_c = \pi a^2/\kappa$. For $\tau/t_r = 0.01$, the optimum condition is obtained at $t_r/t_c = 2 \times 10^{-2}$. In this case, the optimum frequencies are about 200 kHz for copper and 20 kHz for iron, taking the beam radius a to be 0.1 mm. The temperature at $r=0$ must be above the melting point; the necessary power Q must be determined from eqn. (A.17).

In high-frequency pulsed heating by normal solid-state lasers, the heat source is usually considered to be continuous. A critical frequency, above which the source can be considered continuous, has been discussed for the heating of sheet material. For a plate of 20 mm diameter, 1 mm thickness, and thermal diffusivity 100 mm²/s, the critical frequency is 76 kHz. This frequency depends not only on the material properties but also on the dimensions of the sheet [16].

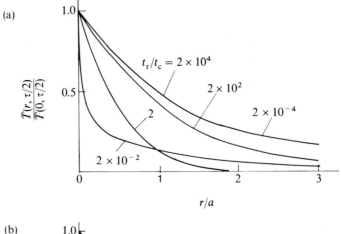

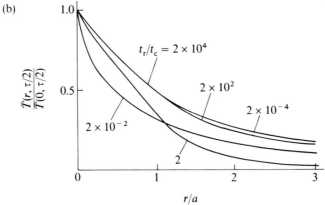

FIG. A.5. Sharpness of temperature distribution [15]. (a) $\tau/t_r = 0.01$. (b) $\tau/t_r = 0.1$

A.6 Effect of variation of thermal properties with temperature

Thermal properties of materials vary with temperature. In the scanning electron or laser-beam annealing of ion-implanted silicon by solid-state epitaxial regrowth, the changes in the thermal conductivity of silicon must be considered, because the conductivity is reduced by a factor of about 5 at the final elevated temperatures [17]. Therefore, a temperature analysis involving thermal properties which vary with temperature is essential in order to obtain the required annealing conditions [17–19].

The basic equation of heat conduction for thermal properties varying with temperature is non-linear, as mentioned in section A.1. Introducing a reduced temperature Θ defined by

$$\Theta = \frac{1}{\lambda_0} \int_{T_0}^{T} \lambda(T) \, dT \tag{A.18}$$

where λ_0 is the thermal conductivity at the reference temperature T_0. Then eqn. (A.1) becomes [20]

$$\frac{1}{\kappa}\frac{\partial \Theta}{\partial t} = \frac{\partial^2 \Theta}{\partial x^2} + \frac{\partial^2 \Theta}{\partial y^2} + \frac{\partial^2 \Theta}{\partial z^2} + \frac{\phi}{\lambda_0} \quad (A.19)$$

In eqn. (A.19), κ is a function of Θ because $\kappa = \lambda/\rho c$. For the steady-state case, $\partial T/\partial t = 0$, eqn. (A.19) obeys the linear equation

$$\frac{\partial^2 \Theta}{\partial x^2} + \frac{\partial^2 \Theta}{\partial y^2} + \frac{\partial^2 \Theta}{\partial z^2} = -\frac{\phi}{\lambda_0} \quad (A.20)$$

When $t \gg t_c$, the temperature rise reaches a steady state. For processing by a moving heat source in which the speed of motion U is so slow that the time determined from a/U or w/U (where a or w represents the incident beam radius) is long compared with t_c, eqn. (A.20) can be adopted as the basic equation.

The relation between reduced linear temperature rise Θ and non-linear temperature rise $T - T_0$ is shown in Fig. A.6 for silicon [17]. It can be seen that there is quite a large difference between the two cases.

For the transient case, numerical calculations are required because eqn. (A.19) is non-linear. For beam annealing of semiconductor devices, considerable care is needed to determine the mesh size in numerical calculations, because of the great difference between beam radius and target dimensions [11].

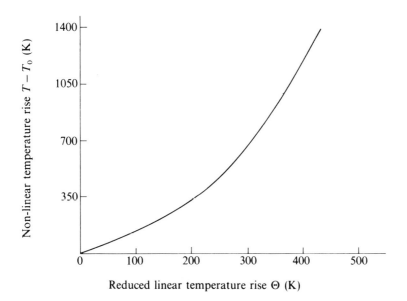

FIG. A.6. Temperature rise involving temperature dependence of thermal conductivity [17]

A.7 Latent heat of fusion and/or vaporization

The mathematical treatment involving latent heat of fusion or vaporization in localized heating is so complex that there have been no analytical solutions so far. Consider one-dimensional heat conduction in a semi-infinite solid workpiece. If a uniform heat flux is applied to the workpiece surface for a sufficiently long time, the surface at $z=0$ reaches the melting point. If the liquid is removed immediately as it is formed, the surface of the workpiece, which was initially at $z=0$, moves inward, and at time t is at position $s(t)$. In this case there is a steady-state solution. For the steady-state rate of melting in terms of the original variables, ds/dt is given [21] by

$$\frac{ds}{dt} \to U = \frac{q}{\rho[L_m + c(T_m - T_0)]} \quad (A.21)$$

where T_0 is the reference temperature. The temperature distribution $T(z, t)$ at time t is given by

$$T \simeq T_0 + (T_m - T_0)\exp\left[-\frac{z - U(t - t_m)}{\kappa}U\right] \quad (A.22)$$

where t_m is the time necessary for generation of melting, and z is the distance from the initial surface at $t=0$. Calculated results of melting rate including those in the unsteady state are shown in Fig. A.7. When the one-dimensional analysis is applied to the drilling, the hole depth D obtained is proportional to the pulse duration τ, because $D = U\tau$ from eqn. (A.21).

In practice, however, analysis involving the latent heat L_m is not simple, because it is not clear whether stock removal is caused by melting, vaporization, or both. When removal is due only to melting, the latent heat of fusion can be adopted. When the

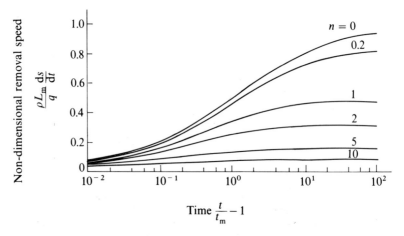

FIG. A.7. Removal speed of melted surface for a semi-infinite workpiece [21].
$$n = \pi^{\frac{1}{2}}c(T_m - T_0)/(2L_m)$$

input heat flux of the beam q has a Gaussian distribution, the power density at the centre of the input heat flux is used as the standard q in general. Furthermore, in laser drilling, the reflection and absorption of the input laser beam by vaporized material just above the working point should be also considered. The tendency for the machined depth to be proportional to the pulse duration agrees well with experimental results for both laser- and electron-beam drilling.

Simple consideration of the latent heat of fusion can be applied, using the equivalent temperature increase given by L_m/c. In the case of direct molten stock removal, the material is considered to be in the solid state until the temperature reaches the adjusted melting point T'_m defined as

$$T'_m = T_m + L_m/c \tag{A.23}$$

In this treatment, of course, change of state of the material is not considered. In vaporization removal, the same treatment may be possible; however, only the latent heat of vaporization is used, because the latent heat of fusion is in general much less than that of vaporization.

A.8 Computer simulation of thermal energy-beam drilling

Drilling by thermal energy-beam is carried out by a sequence of stock removal in the molten and vaporized states. Analytical treatment of the drilled hole shape is difficult, because the shape of the boundary between the molten and solid zones varies with time, and changes in state of the material are also involved.

A simple simulation for obtaining the temperature distribution during drilling has been reported, using a circular heat source which moves perpendicularly to the workpiece surface [22].

Another simulation of drilled hole shape due to stock removal has been carried out on the basis of removal in the molten state. The procedure uses a finite difference method of explicit form with appropriate zoning of the region to be drilled, and the assumption that as the material is melted, it is forced out of the hole by vaporizing material [23]. Calculated examples of hole shape are shown in Fig. A.8, which are in close agreement with experimental results.

A.9 Temperature analysis around the melted zone in welding

The temperature distribution in welding appears more complex than in machining, because the welding involves heat transfer due to convection in the molten metal. The first analysis of the problem used a model with a line heat source in an infinite plate, which is valid for deep penetration and full penetration welding. Although this solution involves an infinite temperature at the centre, the results obtained for the temperature distribution and cooling rate at a point far from the heat source appear to be valid and have been applied successfully to practical problems [24, 25]. The temperature distribution in an infinitely extended plate due to a moving thermal beam with Gaussian power density distribution has also been obtained [2].

APPENDIX A

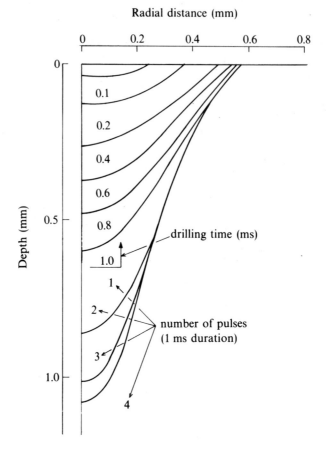

FIG. A.8. Drilled hole shape for molybdenum obtained by simulation [23]. Power $=7$ kW; $w=0.4$ mm

The calculations of temperature distribution above are all based on a given input power. However, there is another procedure, considering the temperature at the boundary between the melted zone and the solid zone to be constant, i.e. at the melting point. If this boundary is assumed to be of a particular shape, the temperature distribution around the melted zone in an infinitely extended plate can be calculated, and thus the heat flow can be obtained from the temperature gradient at this boundary. As losses due to radiation and vaporization are negligible, all the energy deposited in a welding cavity will eventually be lost by conduction to the surrounding solid, because the energy which goes into melting and raising the temperature of the melted zone is returned in the rear part of the melted zone during solidification. Hence the actual input power and temperature profile around the melted zone in welding can be obtained from this calculation.

The boundaries available for this treatment are of three different shapes: circular cylinder, cylinder with forward half semicircular and rear half wedge-shaped, and elliptical cylinder. For the first of these, a solution has been derived [26] and utilized for analysing heat transfer in welding [27]. For the second, the temperature distribution has been calculated by a numerical method [28]. A solution for the third has also been derived and utilized to obtain the heat flow [29].

The actual shape of the melted zone in welding can be treated as an elliptical cylinder. For this shape, the governing temperature equation and boundary conditions in the steady state are given [26] by

$$\frac{\partial^2 T}{\partial x^2} + \frac{\partial^2 T}{\partial y^2} + \frac{U}{\kappa}\frac{\partial T}{\partial x} = 0 \quad (A.24)$$

$$T = T_m \quad \text{at} \quad \frac{x^2}{a^2} + \frac{y^2}{b^2} = 1 \quad (A.25)$$

where a and b are the major and minor axes of the ellipse, respectively, and

$$T \to 0 \quad \text{as} \quad x \to \pm\infty \quad \text{and} \quad y \to \pm\infty \quad (A.26)$$

Introducing elliptical coordinates ξ and η, i.e.

$$x = h \cosh\xi \cos\eta \quad (A.27)$$

$$y = h \sinh\xi \sin\eta \quad (A.28)$$

where $2h$ is the interfocal distance common to a family of confocal ellipses and hyperbolas, the solution of eqns. (A.24)–(A.26) is

$$\frac{T}{T_m} = e^{-2\sqrt{p}\cosh\xi\cos\eta} \sum_{n=0}^{\infty} C_n ce_n(\eta, -p) Fek_n(\xi, -p) \quad (A.29)$$

where non-dimensional parameter p is $(Uh/4\kappa)^2$, $ce_n(\eta, -p)$ is the Mathieu function for the Mathieu equation, $Fek_n(\xi, -p)$ is that for the modified Mathieu equation, and C_n is a constant given by

$$C_{2n} = \frac{2(-1)^n}{Fek_{2n}(\xi_0, -p)} \sum_{r=0}^{\infty} (-1)^r A_{2r}^{(2n)} I_{2r}(\omega) \quad (A.30)$$

$$C_{2n+1} = \frac{2(-1)^n}{Fek_{2n+1}(\xi_0, -p)} \sum_{r=0}^{\infty} (-1)^r B_{2r+1}^{(2n+1)} I_{2r+1}(\omega) \quad (A.31)$$

where $h\cosh\xi_0 = a$, $h\cosh\xi_0 = b$, $\omega = 2\sqrt{p}\cosh\xi_0$, $A_{2r}^{(2n)}$, $B_{2r+1}^{(2n+1)}$ are functions of p and separation constant in solving eqn. (A.24) by eqns. (A.25) and (A.26), and I_r is the modified Bessel function. The temperature distribution for $a/b = 2.0$ is shown in Fig. A.9 [30].

The local heat flux $q(\eta)$ from the cylinder surface at temperature T_m into the surrounding solid is given by differentiation of eqn. (A.29) with respect to the normal to the surface; therefore, the relation for a total heat flow Q_0, melting point T_m, thermal conductivity λ, and thickness H of the cylinder is expressed as

$$\frac{Q_0}{T_m \lambda H} = -2 \int_0^\pi \left.\frac{\partial(T/T_m)}{\partial\xi}\right|_{\xi_0} d\eta \quad (A.32)$$

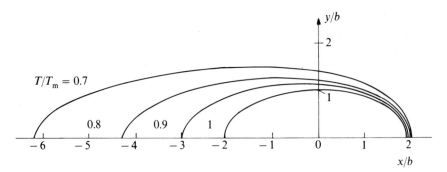

Fig. A.9. Isothermal planes around a moving elliptical cylinder [30]. $\delta = a/b = 2$; $Ub/2\kappa = 1.155$

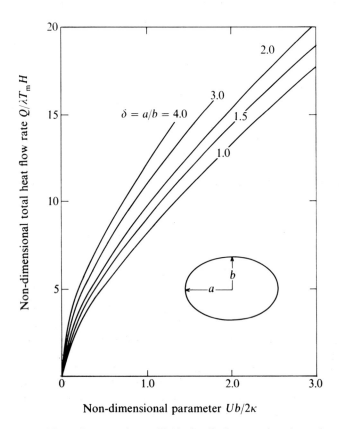

Fig. A.10. Total heat flow rate from elliptical cylinders moving through an infinite plate [29]

Results obtained by numerical integration of eqn. (A.32) are shown in Fig. A.10, where $\delta = a/b$ [29] and the results for $\delta = 1$ is obtained from ref. [26].

The result shown in Fig. A.10 was utilized for the calculation of partial penetration welding shown in Fig. 3.16 (Chapter 3) for electron-beam processing.

In thermal fusion cutting, the existence of a cavity in the plate behind the beam makes mathematical analysis complicated. The temperature distribution has been obtained by introducing a cutting heat source model which is mathematically logical on both walls of the cut zone [31].

A.10 Temperature analysis in the melted zone in welding

Deep but partial penetration welding or full penetration welding in a thin plate can be approximated by steady movement of a cylindrical cavity model surrounded by a thin molten layer through a solid, as shown in Fig. A.11. The cavity, however, is not stable (see section 3.3 on electron-beam processing); in particular, the movement of the molten metal at the rear fluctuates violently. On the other hand, it appears that the molten metal flow at the front is in a stable state; it can be assumed to be in a steady and two-dimensional state.

The temperature distribution, molten metal flow velocity, and molten layer thickness have been obtained under the following assumptions [32, 33]:

 (i) the melted zone is elliptical;
 (ii) the pressure inside the cavity is uniform;
 (iii) the driving force for the molten metal is the surface-tension force due to the temperature gradient on the free surface of the cavity.

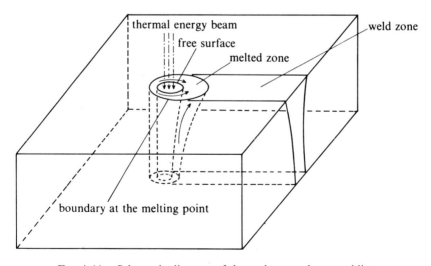

Fig. A.11. Schematic diagram of thermal energy-beam welding

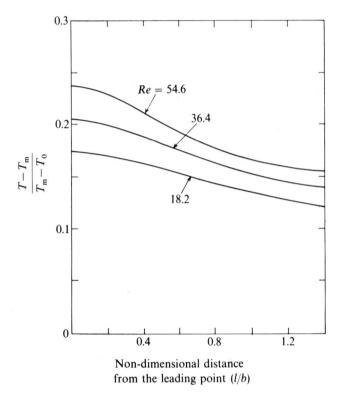

FIG. A.12. Non-dimensional temperature distribution on free surface in the melted zone of a weld [33]

The non-dimensional temperature distribution on the free surface is shown in Fig. A.12 [33], where the abscissa represents a non-dimensional value l/b obtained by dividing the length l along the free surface from the leading point of the cavity by half the melted zone width $2b$. The Reynolds number Re of 36.4 corresponds to welding of aluminium at $U = 18$ mm/s, $b = 1.5$ mm, and kinematic viscosity v (m^2/s) for an approximate average temperature of 973 K. The temperature change from the leading point, $l/b = 0$, to $l/b = 1.4$ is 45 K and the temperature at the leading point is 130 K above the melting point.

References

[1] Carslaw, H. S. and Jaeger, J. C., *Conduction of heat in solids* (2nd edn, Clarendon, Oxford, 1962), 10.
[2] Pittaway, L. G., *Proc. Electron Beam Symp. 5th Annual Meetings* (Alloyd Electronics Corp. Boston, Mass., 1963), 88.

[3] Taniguchi, T. and Yoneyama, T., *Proc. Japan Precision Soc. Meeting* (1966).
[4] Shimizu, R., Ikuta, T., and Murata, K., *J. Appl. Phys.* **43** (1972), 4233.
[5] Cosslett, V. E. and Thomas, R. N., *Brit. J. Appl. Phys.* **15** (1964), 883.
[6] ibid. **16** (1965), 779.
[7] Okabe, S. et al., *J. Japan Soc. Appl. Phys.* **43** (1974), 909 (Japanese).
[8] Kanaya, K., *Bull. Electrotechnical Lab.* **19** (1954), 217.
[9] Dudek, H. J., *Z. angew. Physik* **31** (1971), 331.
[10] ref. 1., 78 and 348.
[11] Iranmanesh, A. A. and Pease, R. F. W., *J. Vac. Sci. Technol.* **B1** (1983), 91.
[12] Lax, M., *J. Appl. Phys.* **48** (1977), 3919.
[13] Miyazaki, T., *Bull. Japan Soc. Prec. Engg.* **13** (1979), 207.
[14] Nagami, H., Suzuki, M., and Katsuta, M., Japan Soc. for The Promotion of Science, 132 Committee, *Report 22* (1961).
[15] Taniguchi, N. and Maezawa, S., *Proc. Electron Beam Symp. 5th Annual Meetings* (Alloyd Electronics Corp. Boston, Mass., 1963), 135.
[16] Rykalin, N. N., Uglov, A. A., and Makarov, N. I., *Soviet Physics-Doklady* **12** (1967), 644.
[17] Lax, M., *Appl. Phys. Lett.* **33** (1978), 786.
[18] Nissim, Y. I., Lietoila, A., Gold, R. B., and Gibbons, J. F., *J. Appl. Phys.* **51** (1980), 274.
[19] Moody, J. E. and Hendel, R. H., *J. Appl. Phys.* **53** (1982), 436.
[20] ref. 1, 11.
[21] Landau, H. G., *Quart. Appl. Math.* **8** (1950), 81.
[22] Paeh, U-C. and Gagliano, F. P., *Trans. IEEE* **QE-8** (1972), 112.
[23] Yoshioka, S. and Miyazaki, T., *J. Japan Soc. Prec. Engg.* **48** (1982), 1028 (Japanese).
[24] Rohsenthal, D., *Trans. ASME* **68** (1946), 849.
[25] Meyer, P. S., Uyehara, O. A., and Borman, G. L., *Welding Research Council Bulletin* **123** (1967).
[26] ref. 1, 390.
[27] Tong, H. and Giedt, W. H., *Trans. ASME. Series C, J. Heat Transfer* **93** (1971), 155.
[28] Arata, Y. and Kanayama, M., *Proc. 2nd Int. Symp. Japan Weld. Soc.* (Osaka, 1975), 21.
[29] Miyazaki, T. and Giedt, W. H., *Int. J. Heat Mass Transfer* **25** (1982), 807.
[30] Miyazaki, T. and Giedt, W. H.: ASME 1985 Winter Annual Meeting, Paper 85-WA/HT-38.
[31] Kanayama, M., Kato, K., and Arata, Y., *J. High Temperature Soc. of Japan* **2** (1976), 322 (Japanese).
[32] Giedt, W. H. and Wei, P. S., *Proc. 7th Int. Heat Transfer Conf.* (Munchen, 1982), 403.
[33] Giedt, W. H. and Wei, P. S., *Modelling of casting and welding processes: Proc. 1983 Engineering Foundation Conf.* (New England College, 1983).

APPENDIX B: MONTE CARLO COMPUTER SIMULATION OF ION-BEAM PROCESSING

B.1 Introduction

The phenomenon of ejection of atoms from a solid target surface subjected to bombardment by energetic ions is known as 'sputtering' (see Fig. B.1). Recently, ion sputter machining has been widely applied in the manufacture of ultra-high-precision parts and ultra-fine parts of electronic, optical, and mechanical devices, because only machining on this atomic scale is capable of making these minute parts with the necessary accuracy.

There have been several analytical studies of sputtering phenomena. These have dealt in particular with the sputtering yield, which is defined as the ratio of the number of sputtered atoms to that of impinging ions. Macroscopic statistical theories have been introduced by Thompson [1], Sigmund [2], and Brand et al. [3] for amorphous and polycrystalline solids, in which the assumption of random retardation of ions in an infinite medium is applied. However, these theories cannot be used effectively for detailed analysis of the sputter machining process, because they are too idealized.

On the other hand, there are the more useful microscopic numerical analyses of Ishitani et al. [4] and others. These use the 'Monte Carlo simulation' technique, based on the LSS theory (Lindhard, Scharff, and Schiott) [5], developed principally for the ion implantation process. These numerical analyses are concerned mainly with the scattering of implanted high-energy ions and recoiled atoms in target materials, but not with the sputtering of target atoms due to impinging low-energy ions of several keV.

In this appendix, to estimate the detailed characteristics of sputter machining with low-energy Ar^+ and Kr^+ ions of about 1 keV, numerical analysis by 'Monte Carlo simulation' is described, with modifications for surface effects.

B.2 Monte Carlo simulation of scattering by ions within target materials

For this calculation, the following assumptions are introduced:

(i) amorphous or polycrystalline target materials
(ii) elastic collision of two bodies
(iii) single scattering
(iv) geometrical flat target surface.

The behaviour of primary impinging (incident) ions and bombarded atoms of the target material can be determined by the four independent parameters shown in Figs. B.2 and B.3, namely: scattering angle, θ, relative to the laboratory frame (or ϕ, relative

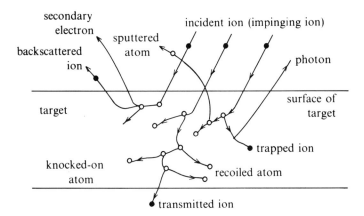

FIG. B.1. Schematic representation of various effects of energetic ion bombardment of solid target

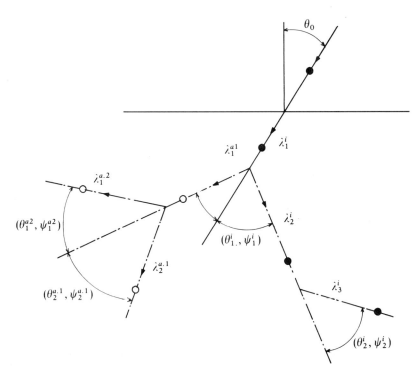

FIG. B.2. Monte Carlo simulation model of retardation process of incident ion in solid material. λ = path length (distance between one collision and the next); θ = scattering angle with respect to trajectory before the collision; ψ = azimuthal scattering angle with respect to trajectory before collision

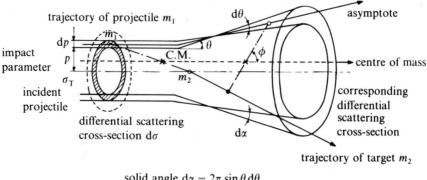

FIG. B.3. Elementary parameters for classical single scattering problem. $\delta_t =$ maximum total scattering cross-section $(A) = \pi (l/2)^2$; $l =$ mean lattice constant

to the centre of the mass frame); azimuthal scattering angle, ψ, energy of impinging ion, E_i; and path or step length, λ (distance between one collision and the next). These four parameters governing the behaviour of the ion or the atoms of the target material are basically determined by random numbers generated by digital computer. The relation between the scattering angles θ and ϕ is determined by classical dynamic collision theory as follows:

$$\cos \theta = (M_i + M_t \cos \phi)/(M_i^2 + 2M_i M_t \cos \phi + M_t^2)^{\frac{1}{2}} \tag{B.1}$$

where M_i is the mass of the impinging ion, and M_t is the mass of target atom at rest.

B.2.1 Determination of scattering angle θ or ϕ

It is verifiable by classical elastic atomic collision theory that an impinging ion entering the differential scattering area $d\sigma$ is always scattered at the corresponding scattering angle between θ and $\theta + d\theta$, or ϕ and $\phi + d\phi$, which is related to $d\sigma$ by the following equation, using the parameter t instead of θ or ϕ:

$$d\sigma = a^2 (2t)^{\frac{3}{2}} f(t^{\frac{1}{2}}) dt \tag{B.2}$$

$$f(t^{\frac{1}{2}}) = \lambda t^{\frac{1}{2}-m} [1 + (2\lambda t^{1-m})^b]^{-\frac{1}{b}} \tag{B.3}$$

where $t^{\frac{1}{2}} = \varepsilon \sin(\phi/2)$, the normalized transferred energy; $\varepsilon = a/b$, the normalized impinging energy of the ion; a is the screening radius (m); $b = Z_i Z_t e^2/(4\pi \varepsilon_0 E_r)$ (m), the shortest distance between two particles in collision at the impinging energy E_r (J); $E_r = (\mu/M_i) E_i$, the energy of the impinging particle with reference to the relative frame (J), corresponding to the impinging energy of the ion E_i (J), with reference to the laboratory frame; $\mu = (M_i M_t / M_i + M_t)$, the reduced mass with reference to the relative frame (kg); M_i, M_t are the masses of the ion and target atom respectively (kg); Z_i, Z_t are the atomic numbers of the ion and target atom at rest respectively; e is the electronic charge, 1.603×10^{-19} C; and ε_0 is the dielectric constant of vacuum, 8.8542×10^{-12} F/m.

The term $f(t^{1/2})$ is called the scaling factor of screening potential, at a Thomas–Fermi potential field for impinging ions of higher energy than E^*, where E^* is the critical energy of impinging ions at the boundary of the range of the Born–Mayer potential field. For Ar^+ ions and Si atoms, E^* is 336 eV, and values of the indices m, q, and λ are given in Table B.1. For example, $f(t^{1/2})$ for argon ions vs. silicon atoms is shown in Fig. B.4.

At a Born–Mayer potential field for impinging ions of lower energy than E^*, the scaling factor is reduced to the form

$$f(t^{\frac{1}{2}}) = \lambda t^{\frac{1}{2} - m} \tag{B.4}$$

TABLE B.1
Values of m, q and λ

Potential	m	q	λ
Thomas–Fermi	0.333	0.667	1.309
Born–Mayer	0	–	24

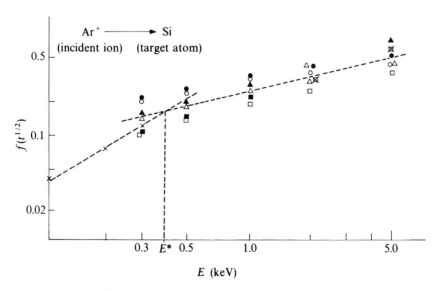

FIG. B.4. Scaling factor of screening potential

Consequently, using eqn. (B.2), the probability $p(\phi)$ that the incident ion is scattered at an angle between ϕ and $\phi + d\phi$ can be defined by eqn. (B.5) (see Fig. B.2):

$$p(\phi)d\phi = d\sigma/\sigma_T \qquad (B.5)$$

where σ_T is the total scattering cross-section, which means the effective cross-sectional area for collision of the incoming ion with the atom at rest. The quantity $d\sigma/\sigma_T$ is determined by the random probability rule, because the impinging ion is randomly projected into $d\sigma$. Accordingly, the following relation should be realized when the uniformly distributed random number η_i is defined as between 0 and 1:

$$\eta_i = \int_0^\sigma \frac{d\sigma}{\sigma_T} = \int_{\phi_{min}}^\phi p(\phi)d\phi \qquad (B.6)$$

where, $\phi = \phi_{min}$ at $\eta_i = 0$ and $\phi = \pi$ at $\eta_i = 1$, and at the Thomas–Fermi potential field, the least scattering angle $\phi_{min} = 2$.

On the other hand, σ_T is defined as follows:

$$\sigma_T = \int_{\phi_{min}}^\pi (d\sigma/d\phi)d\phi \qquad (B.7)$$

B.2.2 Determination of energy transferred to bombarded target atom, E_a

If the scattering angle ϕ of an incident ion is determined, as mentioned above, by the elastic collision theory, the replaced energy, E_a, of the atom at rest can be determined as follows:

$$E_a = 4M_i M_t/(M_i + M_t)^2 E_i \sin^2(\phi/2) \qquad (B.8)$$

B.2.3 Determination of path length λ

From the Poisson distribution, the path length λ is determined by a random number γ_i:

$$\lambda = -L \ln \gamma_i \qquad (B.9)$$

where $L = 1/(N\sigma_T)$, and N is the number of atoms per unit volume of target material.

B.2.4 Determination of azimuthal scattering angle ψ

It is naturally considered that the azimuthal scattering angle ψ should be distributed randomly within the range 0 to 2π.

The four parameters determining the behaviour of ions and atoms in the target after collision can be determined from the random numbers generated by a digital computer. The Monte Carlo simulation has been performed as shown by the flow chart in Fig. B.5. Two examples of trajectories obtained for incident ions and recoiled atoms are shown in Figs. B.6 and B.7.

B.3 Modifications to simulation model for collision of ions of lower energy

B.3.1 Determination of least scattering angle ϕ_{min} as a function of incident ion energy

If ϕ_{min} is given the constant value of 2°, the total scattering cross-section at σ_T calculated from eqn. (B.7) is larger than the mean cross-sectional area of the target

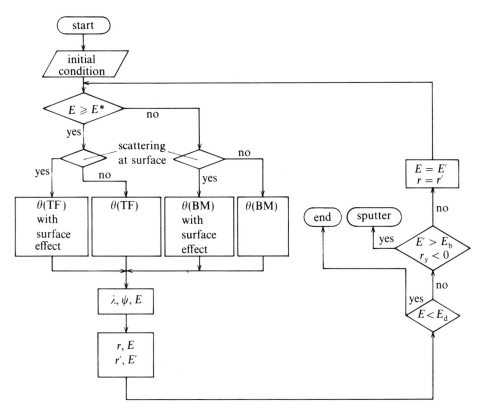

FIG. B.5. Flow-chart of Monte Carlo computer simulation. θ = scattering angle; λ = path length; ΔE = energy loss; ψ = azimuthal scattering angle; r, r^1 = position vectors of projectile and target; E, E^1 = energy of projectile and target; E_b = surface barrier energy; E_d = displacement energy; E^* = critical energy; BM = Born–Mayer; TM = Thomas–Fermi

atom $\pi(l/2)^2$, where l is the mean distance between randomly located target atoms. Therefore, by controlling the value of ϕ_{min}, σ_T can be determined so as not to exceed the value of $\pi(l/2)^2$, that is, contrary to the Thomas–Fermi potential field, ϕ_{min} is determined by the following equation:

$$\sigma_T = \int_{\phi_{min}}^{\pi} (d\sigma/d\phi)d\phi = \pi(l/2)^2 \tag{B.10}$$

From this modification to ϕ_{min}, the scattering angle θ for incident ions of lower energy can be determined as a function of the random number.

B.3.2 Surface effects for incident ion beam and sputtered atoms

The impact parameter or the scattering angle and the azimuthal scattering angle due to surface atoms are largely restricted by the incident angle of low-energy ions.

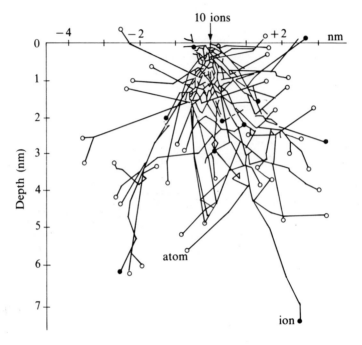

FIG. B.6. Trajectories of incident ions and recoiled atoms in target: example (1) (10 ions). $Ar^+ \to Si$. Ion incident energy, $E = 1.0$ keV; ion incident angle, $\theta = 0°$

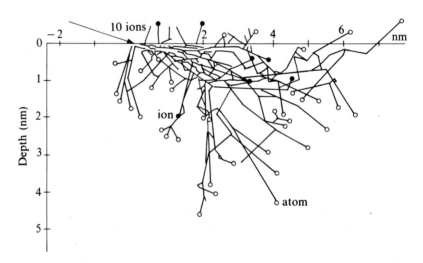

FIG. B.7. Trajectories of incident ions and recoiled atoms in target: example (2) (10 ions). $Ar^+ \to Si$. $E = 1.0$ keV; $\theta = 75°$

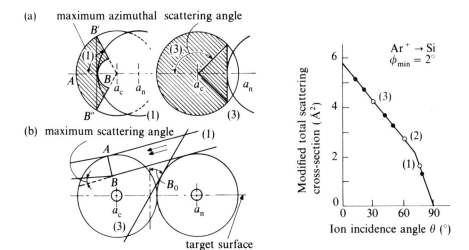

FIG. B.8. Simple model of surface masking ($R = l/2$)

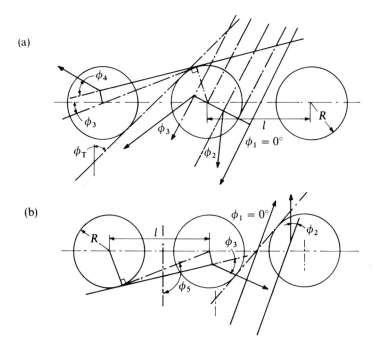

FIG. B.9. Surface effects for (**a**) incident ion ($R < l/2$), and (**b**) sputtering ($R < l/2$)

Incoming ions are masked at the surface by neighbouring atoms. To allow for this masking effect in the simulation, simplified models for incident ions and sputtered atoms are proposed, as shown in Fig. B.8 and Fig. B.9. For a total scattering area σ_T of $(l/2)^2$, the incident ions within the shaded sector shown in Fig. B.8 are influenced by surface atoms. That is, the impact parameter, or scattering angle, due to surface atoms is confined within sector 3 when the incidence angle is less than 60°, and is confined within sector 1 when the incidence angle is larger than 60°. The scattering angle θ_0 is confined within the range 61.2° to −61.2°, where, in the case of $E_i > E^*$,

$$\phi_{max} = 2\sin^{-1}[(2R^2 m/\lambda_{TF} a^2 \varepsilon^2) + 1]^{-\frac{1}{2}m} \qquad (B.11)$$

and in the case of $E_i < E^*$,

$$\phi_{max} = 2\sin^{-1}[\exp(-R^2/\lambda_{BM} a^2)] \qquad (B.12)$$

with $\quad R = r - 2r\cos\theta_0, \quad r = l/2, \quad$ and $\quad m = 0.333, \quad \lambda_{TF} = 1.309$

(as given in Table·B.1), the subscripts BM and TF denoting Born–Mayer and Thomas–Fermi respectively.

On the other hand, when $\sigma_T < \pi(l/2)^2$, the same masking model is proposed to take effect in the range of incidence angle greater than $\theta_s = \cos^{-1}(2R'/L)$. Furthermore, in the case of the impact parameter corresponding to the gap of two total scattering area, the scattering angle of incident ions is 0° (Fig. B.9). The surface effects of sputtering may also be deduced using the same kind of model.

B.4 Adaptation of the Kinchin–Pease model

The defects in the target material created by the collision of incident ions with atoms at rest can be classified by the Kinchin–Pease model as shown in Table B.2, by which the creation of vacancies, interstitial atoms, and substitutional atoms can be easily identified.

TABLE B.2
Diagrammatic representation of Kinchin–Pease model for the creation of lattice defects

Energy of struck atom	(E'_2)	no defect	vacancy	Energy of struck atom	(E'_2)	ion substituted for Si	vacancy
	E_d	interstitial atom	no defect		E_d	interstitial ion	no defect
		E_d	(E'_1)			E_d	(E'_1)
		Energy of recoiled atom				Energy of incident ion	

B.5 Results of Monte Carlo simulation and discussion of experimental data

An experiment was performed with an ion-beam sputter-machining apparatus of the ion-shower type or Kaufman type as shown in Fig. 1.25 (Chapter 1). To obtain experimental data on the angular distribution of sputtered atoms, a collimator and a collector plate (glass) were arranged as shown in Fig. B.10.

(a) Figures B.11 and B.12 show the calculated and experimental data on sputtering yield vs. energy of incident Ar^+ and Kr^+ ions for the (100) surface of an Si crystal. In both these cases, the experimental curves lie above those calculated by Monte Carlo simulation. The reason for this seems to be the increase in sputtering yield due to the temperature rise of 200–300°C in the workpiece caused by the ion bombardment. Increasing the temperature by 300°C causes a 20% increase in sputtering yield compared with that at absolute zero, at which the Monte Carlo simulation is performed. The corrected data are fairly well fitted to the experimental data above an energy of 1 keV, but not below this. The reason for this seems to be that the assumption of Born–Mayer potential at lower energy is inappropriate.

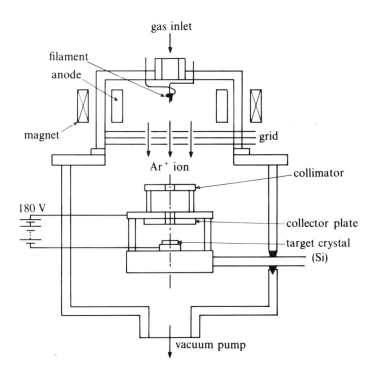

FIG. B.10. Experimental set-up for obtaining the distribution of sputtered atoms

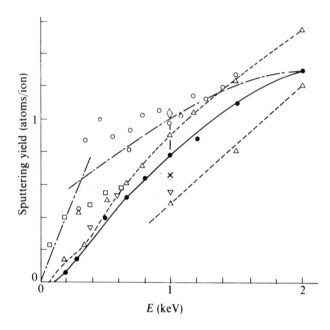

- ● : estimated by Monte Carlo
- ▲ : estimated (300°C)
- —·—: theoretical (TF by P. Sigmund)
- -----: theoretical (BM by P. Sigmund)
- ○ : experiment
- △ : experiment (Southern)
- □ : experiment (Wehner)
- × (100)
- ◊ (111) } experiment (Hosaka and Hashimoto)
- ▽ : experiment (Carter and Colligon)

FIG. B.11. Dependence of sputtering yield on ion acceleration energy. $Ar^+ \rightarrow Si$; $\theta = 0°$

(b) Figures B.13 and B.14 show the calculated data on sputtering yield vs. incidence angle of Ar^+ and Kr^+ ions respectively. Both sets of experimental data are fairly well fitted to the results of Monte Carlo simulation, but effects due to the temperature rise still exist.

(c) The calculated and experimental data on the solid angle distributions of atoms sputtered from an Si target by Ar^+ ions for $E = 1$ keV, $\theta = 0°$, are shown in Fig. B.15. The experimental pattern almost follows a cosine distribution. On the other hand, the results obtained by Monte Carlo simulation, assuming no surface effects follow a sub-cosine curve. When the surface effects shown in Fig. B.9 are taken into

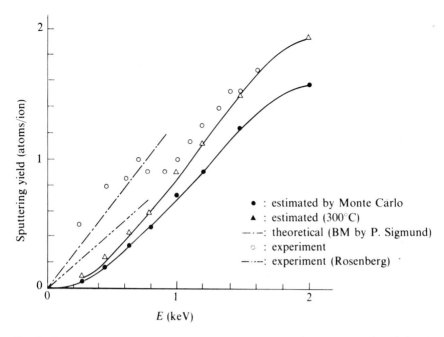

FIG. B.12. Dependence of sputtering yield on ion acceleration energy. $Kr^+ \to Si$; $\theta = 0°$

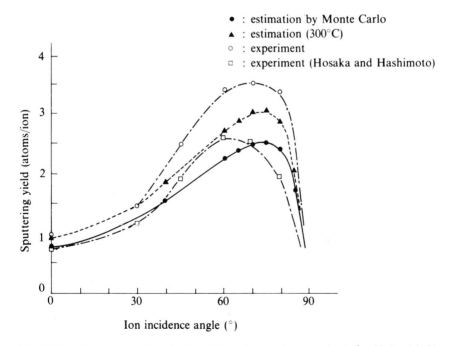

FIG. B.13. Dependence of sputtering yield on ion incidence angle. $Ar^+ \to Si$; $E = 1$ keV

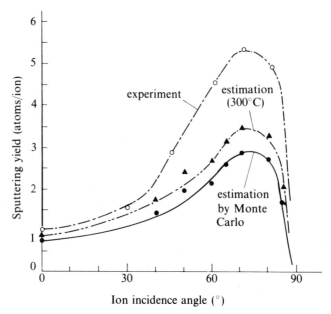

FIG. B.14. Dependence of sputtering yield on ion incidence angle. $Kr^+ \to Si; E = 1$ keV

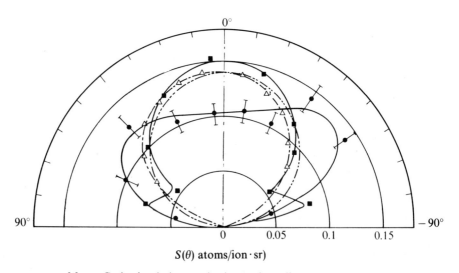

- ● : Monte Carlo simulation, neglecting surface effect
- ■ : Monte Carlo simulation, including surface effect
- △ : Experiment
- —·—: Simulation and experiment (10 keV, Suk and Shimizu) (scale: 1/30)

FIG. B.15. Distribution of ejected atoms. $Ar^+ \to Si; E = 1$ keV; $\theta = 0°$

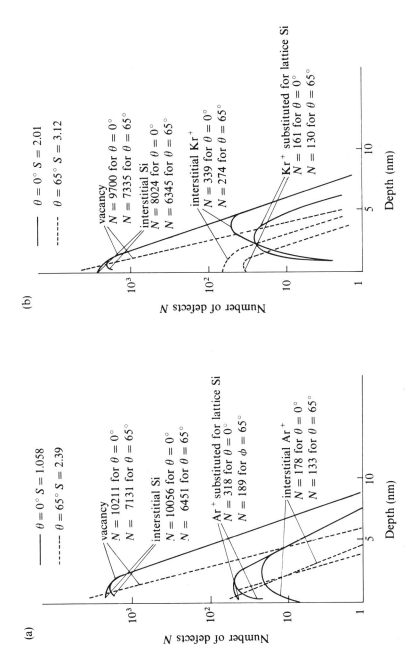

FIG. B.16. Depth distributions of four types of crystal defect for 500 incident ions at 1 keV (**a**) Ar$^+$→Si. (**b**) Kr$^+$→Si.

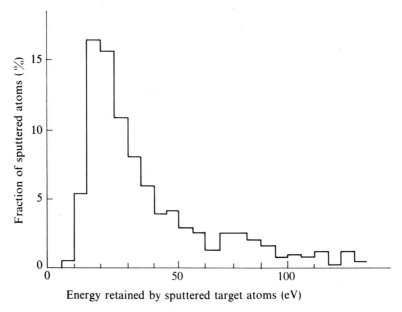

FIG. B.17. Energy distribution of sputtered target atoms. $Ar^+ \rightarrow Si$. $E = 1.0$ keV; $\theta = 0°$

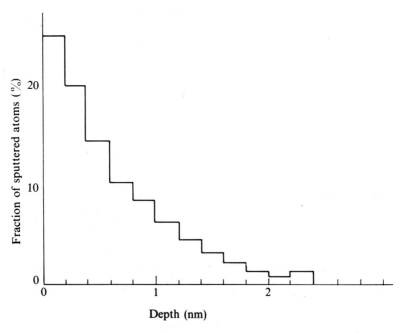

FIG. B.18. Original depth distribution of sputtered target atoms. $Ar^+ \rightarrow Si$. $E = 1.0$ keV; $\theta = 0°$

account, the calculated results start to resemble closely the experimental data. For incidence angles of 0°, 40°, and 70°, calculated and experimental data are shown in Fig. B.11. Considering the calculated result for the incidence angle of 40°, the surface effect seems to be overestimated.

(d) The depth distribution of lattice defects of four types, that is, vacancy, interstitial silicon atom, interstitial Ar^+ or Kr^+ ion, and substitutional Ar^+ or Kr^+ ion at Si crystal site, is shown in Fig. B.16. Comparing the defects caused by Ar^+ and Kr^+ ions, the distributions of interstitial ions differ considerably, but those of other defects are nearly the same. Moreover, vacancies and corresponding interstitial silicon atoms are fairly large and nearly the same in number, but as they are paired, the Si atoms seem to recover their positions in the crystal sites rapidly at room temperature.

(e) The energy distribution of sputtered atoms is shown in Fig. B.17, and the initial depth distribution of sputtered atoms in Fig. B.18. These data show that more than 90% of sputtered atoms have an energy below 80 eV, and more than 98% atoms are sputtered out from the surface layer within a depth of 1 nm.

B.6 Conclusions

The results of calculations based on Monte Carlo simulation with two kinds of modified model, surface masking and Kinchin–Pease, fit the experimental data obtained by the authors fairly well, but to obtain better correspondence, it seems to be necessary to correct the models. This kind of microscopic simulation may be useful in fundamental studies of ion sputter machining, but for practical use, macroscopic statistical methods may be more convenient for calculating the sputtering yield and other macroscopic machining characteristics.

References

[1] Thompson, M. W., *Phil. Mag.* **18** (1968), 377.
[2] Sigmund, P., *Phys. Rev.* **187** (1969), 383.
[3] Brand, W. et al., *Nucl. Inst. Meth.* **47** (1967), 201.
[4] Ishitani, I. et al., *Jpn. J. Appl. Phys.* **11** (1972), 125.
[5] Lindhard, J. et al., *Kgl. Danske Videnskab. Selskab. Mat.-Phy. Medd.* **36** (1968), 10.

INDEX

aberration 90
absorption 64
   coefficient of 70
   increases in 68
accuracy 44
acoustic-wave filters 237–9
activated reactive ion plating (ARIP) 266
activation energy 45, 50
active defects 9
adjust melting point 287
alexandrite 77
alloying 127–8
aluminium, etching 232–4
anisotropic etching 226
annealing 52, 126–7, 138
   electron beam 177–8
Arrhenius' reaction rate theory 50, 217
aspheric lens-making 249–51
atomic bonding energy 45
atomic energy 4, 8–9
atomic energy-beam processing 45–7
atomic lattice range threshold 15–17
atomic lattice scale 10
atomic-scale deposition 60
atomic-scale matching 41–4
atomic surface barrier energy 16
atoms
   and the Kinchin–Pease model 302
   surface 32
austenite 75
azimuthal scattering angle 298

balancing 112
beam waist, CO_2 lasers 89
   measurement of 100
biaxial CO_2 laser 84
Boltzmann theory 50
bombarding energetic ion 17
bonding 191
Born–Meyer potential 297
bubble splashing 32–3

carbon content 67
carbon dioxide lasers 79, 82–7
   biaxial type 84
   coaxial type 83–4
   cross-flow type 83
   optical components of 93–6
   orthogonal triaxial type 85–7

   reflectance, with metal mirrors 94
   transparent components in 93–5
   triaxial type 83, 85
CCT 180
ceramic welding 118–24
ceramics, and metal eutectic 129–30
characteristic response time 21, 283
chemical analysis, ion-beam 271–2
chemical/electrochemical reactants 49
chemical etching 29
chemical etching, ion-beam assisted
   (IBAE) 222–4
chemical process 49–52
coaxial CO_2 laser 83–4
computer simulations
   Monte Carlo, ion-beam
     processing 294–309
   thermal energy-beam drilling 287
conical holes 102
cross-flow type CO_2 lasers 83
crystal growth 46, 133
crystal growth semi-conductors 30
curtain beam 174
cutting
   ability 107
   and electron beams 149
   lasers 109, 129
   scribing 103–12
   shear stress 14
   speed, lasers 109
   under- 251

d.c. glow discharge 82
deep-penetration weld 32, 34
degradation 175
depolymerization 138
diamond tools 247–8
diffusion, of internal energy 19
dislocation–microcrack scale 10–21
dislocation range 59
divergence angle 90
doping 53
drilling
   electron beam 142–9
   laser beam 101–3
   thermal energy beam 287
driving potential 49
dross 107
DSR 55

dual ion-beam deposition 261–2
dynamic processing 38

electric capacitance type video discs 55–9
electric discharges 30, 76
electric spark machining of dielectric insulators 38
electrical conductivity 64
electrochemical process 49–52
electro-elastic collision 44–5
electrohydrodynamic ion sources (EHDIS) 43
electron-beam annealing 177–8
  scanning 177–8
  semi-conductors 177
electron-beam curing 176
electron-beam heat treatment 179–82
  hardening 179–80
  melting 180–2
  surface modification 182
electron-beam image projection 171–2
electron-beam lithography 35, 162–72
  equipment 164–7
  limitations 167–72
  process of 163–4
electron-beam machining 142–50
  drilling 142–9
  focusing 143
  foil 149–50
  plate cutting 149
  pulse conditions 144–9
electron-beam melting 187–90
  electron guns and 188–90
  equipment for 187–8
  purification 188
electron-beam processing 138–99
  atomic/molecule 45–7
  basic concept of 1
  chemical/electrochemical 49
  development 138
  electrons 30–8
  equipment for 141–2
  fundamentals of 49–51
  ion beam 38–45
  lithography 162–72
  machining 142–50
  melting 187–90
  micro-fabrication techniques 52–9
  MOS IC 52–5
  need for 1
  photons and 24
  plasma 47–8
  reactive processing 141
  semi-conductors, annealing 177–9
  temperature analysis, thermal energy processing 276–93
  thermal processing 139, 179–82

vapour deposition 182–7
video discs 55, 57, 59
electron beams 76
  reactive processing 34–6
electron-beam shape 166
electron-beam vapour deposition 182–7
  alloys 186–7
  applications of 186
  evaporation and 182–5, 192
  film thickness and 185–6
electron-beam welding 150–61
  applications of 616
  equipment for 159–60
  partial-vacuum 159–60
  penetration mechanism 150–6
  scanning 167
  vacuum types 159–60
  weld zone and 156–9
electron bombardment 138
electron deflection 165
electron-discharge beam processing 36–8
electron-discharge machining 37–8
electron–electron interactions 170–1
electron guns 184, 188, 190
electron penetration 281
  range 32
electron reactive processing 172–92
  applications of 176
  degradation 175
  factors affecting 173–6
  G value 173
  irradiation energy 173
  polymerization and 175–6
  reactions, combination type 174–5
electrons 30–8
  proximity effects of 167–70
electron sources 164–5
energy, *see specific forms of*
etching
  of Al, 232–4
  heat-less 133
  ion beam 216–34
  ion-beam sputter 204–16
  isotopic 204
  lasers 132–3
  maskless 204
  selectivity 226
  of Si 226–7
  of Si/SO$_2$, with CF$_4$/I$_2$ 227, 229
  SiO$_2$ 229–32
etching rate
  gas pressure 221–2
  ion current density 219
  ion energy 217
  ion incidence angle 219
  time, and depth 222
evaporation, and deposition 182–3, 192

INDEX

excimer-laser-beam 55
exposure 52

Fabry–Perot resonator 76
facet formation 251
failure 9–11, 14–15
flexible manufacturing system (LFMS) 135
floating zone melting 191–2
focused-beam direct ion implantation 243
focused ion-beam processing 202–4
focusing, of laser beams 87–91
foil bonding 191
foil, machining of 149–50
fracture 9–11, 15
free energy 4, 8–9

gauge blocks 246–7
Gaussian distribution 278–9
Gaussian round electron beams 166
Gibbs's free energy 4, 8–9, 17, 49
glassy structure surface layers 128
glazing 29, 128
Gnanamuthu classification 124–5
gold plating 130
grain–void scale 12
gratings 240, 242
G value 173

Hagen–Ruben relation 64
hardening 126–7, 179–80
hardness 74
Helmholtz free energy 8–9
holography 29

IBAE 222–4
IC transistor 52
IC wafers 29
image projection, *see* electron-beam image projection
implantation, ion-beam 269–70
input power density 18
input thermal power density 21
internal elastic strain energy 4
internal energy, diffusion 19
inverse population 76
ion-assisted chemical etching 44
ion-beam deposition 253–68
ion-beam direct deposition 268
ion-beam etching 204–6, 216–34
   rate of, and energy 217
   reactive 216–17, 224–6
ion-beam lithography 271
ion-beam processing 38–45, 200–75
   chemical analysis 271–2
   concepts 200
   equipment for 200–4
   ion-beam deposition 253–68

ion-beam lithography 271
ion-beam removal 204–53
ion-beam surface treatment 268–71
Monte Carlo computer simulation 294–309
   power density 44–5
ion-beam removal process 204–53
   problems with 251–3
ion-beam sputter machining 204–16
ion-beam surface treatment 268–71
   implantation 269–70
   mixing 270–1
   nitriding 268–9
ion collision 207
ion deposition 44
ion implantation 39
ion micropore mass analysis 272
ion mixing 44
ion nitriding 44, 268–9
ion plating 44
   reactive 266
ion rubbing 44
ion-shower beam processing 202
ion species 207
ion sputtering 17, 38–45, 204–16
   and atomic-scale machining 41–4
   deposition 253–61
   machining 39–41
isotropic etching 204

Jaeger's formula 69

kerf width 107, 109
Kinchin–Pease model 302
kinetic energy yield 208
Knoop hardness 130–1

Landau's equation 72
lanthanum hexaboride 165
large-scale integrated circuits (LSIs) 29
laser-assisted cutting 129
laser-beam machining
   cutting/scribing 103–12
   drilling 101–3
   micro-machining 112–14
laser-beam processing 62–75
   equipment for 75–100
   kinds of 63–4
   machining 101–14
   melting and 71–2
   residual strain 72–5
   vaporization and 71–2
laser beams 24–30
   beam mode measurement 97–100
   beam waist measurement 100
   focusing 87–91
   interactions 64–8
   light in 62–3

reactive processing 29
surface treatment with 124–8
transmission 91–3
laser coating 129–30
laser-enhanced etching/plating 132–3
laser-enhanced gold plating 130
laser medium 76
laser oscillator 75–7
laser photolysis 132
laser resonator mode 28
lasers
 carbon dioxide 79, 82–7
 YAG 78–9
laser shaping 130, 132
latent heat of fusion/vaporization 286–7
latent image 29
lattice bonding energy 4, 8–9
light, focused, *see* laser beams, *aspects of*
line heat source 287
linear kinetic energy 8
liquid-metal ion sources (LMIS) 43
lithography, *see* ion-beam lithography

machining speed 18
macroscopic density 45
macroscopic input power density 60
magnetic bubble memories 235–6
magnetic ion separation 43
magnetron sputtering 254–61
marking 112
martensite 75
mask repair 243–5
maximum absorption rate 32
Maxwell's energy-distribution theory 50
mechanical processing
 energy-beam processing and 4
 threshold for 11–15
melted zone 291
 in welding 287–91
melting, electron-beam 180–2, 187–90
 floating zone 191–2
 point, adjusted 287
 powder production 192
 surface 181
metallographic transformation 29
metal mirrors 94
metal welding 114–18
microbridge 243
microcrack range 59
micro-fabrication technique 52–9
 MOS IS 52–5
 video discs 55, 57, 59
microfine coining 55
microfine structures 52
micro-ion lithography 55
micro-ion beams 43, 55
microlenses 242

micro-machining, lasers 112–14
microscopic atomic input power 52
microwaves 38
micro-welding 124
mirrors, metal 94
mixing, ion-beam 270–1
molar activation energy 49
molecule energy-beam processing 45–7
Monte Carlo computer simulation 294–309
MOS IC 52–5
multimode beams, lasers 90

nano-technology 44
nascent state of reactive ions 45
nitriding, ion 268–9
nuclear stopping power 206

optical fibres 92
optical resonator 83
optical type video discs 55–9
orthogonal triaxial CO_2 laser 85–7
oxide films 65, 67, 75

parabolic mirrors 91
path length 298
penetration depth 38–9
penetration, and electron-beam welding
 depth of 152–5
 formula for depth 155
 mechanism of 150–2
penetration path 32
penetration welding 291–2
percussion hardening 29
photo-beam processing 62–137
 equipment for 75–100
 fundamentals of 62–75
 laser-beam machining 101–14
 new applications of 129–33
 surface treatment 124–8
 welding/soldering 114–24
photocatalysis 132
photocathode electron image
 projection 171–2
photons 24
photoreactive materials 29
photoresist materials 29
Pittaway equation 69
plasma-arc beam 47
plasma energy-beam processing 47–8
plasma fume 114
plasma gun 47
plasma-jet beam processing 47
plastic slip failure 11, 14–15
plating, ion 262–7
plating, and lasers 132–3
 gold 130
point-defect range 17

point-defect scale 10
Poisson distribution 298
polymerization 138, 175–6
powder production, by melting 192
  evaporation 192
power density 17
  input 51–2
  ion-beam processing 44–5
processing unit size 9
proximity effect 167–71
pulse beams, and heat zone 281–3
pulse duration, and electron beams 144–7
pulse electron-beam annealing 177
pyrolytic etch process 132

quartz oscillator 113

random number 298
random-walk displacement 102
raster scanning 167
reaction rate 50
reactive ion etching 204
reactive processing
  electron 172–92
  electron-beam 141
  energy-beam, fundamentals 49–51
  focused broad ion beam 45
  focused electron beam 34–6
  lasers 29
  radiation 29
reactive radicals 225
redeposition 251
reflectance 64–5
repair 112
residual strain 72–3
resolution 44
r.f. plasma ion-beam processing 201–2
RIE 224–6
ruby 77
Rutherford back scattering (RBS) 272

sapphire crystal 133
saturation temperature 21
scanning 167
  raster 167
  vector 167
scanning electron-beam annealing 177–8
screening potential 297
secondary ion mass spectrometry 272
self-quenching 179–81
semi-conductors 177
  crystal growth 30
semi-infinite solid 281, 283, 286
sensitivity 44
servo systems 44
shaped electron beam 166
shock hardening 29

shear strength 14
shear stress 14
silicon etching 226–32
soldering, see under welding
solids, internal energy diffusion 19
solid-state electronic devices,
  fabrication 234–5
SOR (synchrotron orbit radiation) 24, 29
specific bonding energy 16
specific lattice bonding energy 16
specific processing energy 4–11
specific sputter machining rate 213
specific volumetric lattice bonding energy 16
specific volumetric stock removal energy 15
spinning rotor 113
spot size, CO_2 lasers 89
sputter, ion 204–16
sputter machining rate 210–14
  specific 213–14
sputtered atoms, distribution 214–16
sputtering 295
sputtering, magnetron 254–61
sputtering yield 206–13
stamping process 57–9
statistical electron penetration depth 32
steady-state solution 286
steam turbines 113
strain, residual 72–5
stripping 52
super-lattice substance 46–7
surface acoustic wave filters 237–9
surface barrier energy 4, 8–9
surface energy density 16
surface heat source 277–80
surface polishing 190–1
surface roughening 248–9
surface roughness 68
surface treatment, with lasers 124–8

temperature
  melted zone 291–2
  and thermal properties 284–6
temperature analysis 19–21
temperature gradient 21
temperature rise 66, 277–81
  and laser beams 63, 66, 68–71
  lasers, distribution 69
tensile crack fracture 15
theoretical shear strength 14
thermal conductivity 21
thermal diffusivity 21, 70
thermal energy 4
thermal energy-beam drilling 287
thermal expansion coefficient 70
thermal processing
  by electron beams 30–3, 139
  by focused light 24–9

by heat conduction 277
  temperature 22-3
thermal vibration 17
thermally-assisted field emission 165
thin films, machining 149-50
Thomas–Fermi potential field 297
thoriated tungsten filaments 164
threshold for energy-beam processing 15-17
threshold for point-defect range 17
threshold specific processing energy 9, 16
  estimating 11-17
  and plastic slip failure 11, 14-15
  and tensile crack fracture 15
total scattering cross-section 298-9
transformation hardening 179
transmission, of laser beams 91-3
triaxial CO_2 lasers 83
trimming 112
tuning 112
turning, laser-assisted 129

undercutting 251

vacuum electron-beam welding 159-60
  high- 159-60
  non- 159-60
  partial 159-60

Van der Waals atomic radius 45
vapour bubble 32
vector scanning 167
Vickers microhardness tester 74
video disc masters 29
video discs 55, 57, 59

wafer patterning 52-4
welding, and lasers 114-24
  ceramics 118-24
  metals 114-18
  micro- 124
welding, melted zone 287-92
welding speed, of lasers 114
weld zone, electron-beam processing 156-61
  defects 158-9
  mechanical properties 157
wire bonding 191
WRES 55

X-rays 24, 29
X-ray topography 72

YAG laser 77-9

zone plates 239-40
zoom electron gun, for welding 34